The Reference Shelf®

Global Climate Change

Edited by Paul McCaffrey

The Reference Shelf
Volume 78 • Number 1

The H. W. Wilson Company
2006

The Reference Shelf

The books in this series contain reprints of articles, excerpts from books, addresses on current issues, and studies of social trends in the United States and other countries. There are six separately bound numbers in each volume, all of which are usually published in the same calendar year. Numbers one through five are each devoted to a single subject, providing background information and discussion from various points of view and concluding with a subject index and comprehensive bibliography that lists books, pamphlets, and abstracts of additional articles on the subject. The final number of each volume is a collection of recent speeches, and it contains a cumulative speaker index. Books in the series may be purchased individually or on subscription.

Library of Congress has cataloged this title as follows:

Global climate change / edited by Paul McCaffrey
    p. cm.—(The reference shelf ; v. 78, no. 1)
  ISBN 0-8242-1059-X (alk. paper)
  1. Climatic changes. 2. Global warming. I. McCaffrey, Paul, 1977– II. Series.
    QC981.8.C5G644 2006
    363.738'74—dc22

                    2005032763

Cover: An iceberg in Columbia Bay flows from Columbia Glacier, on April 6, 2004, near Valdez Alaska. (Photo by David McNew/Getty Images)

Visit H.W. Wilson's Web site: www.hwwilson.com

Printed in the United States of America

Contents

Preface

Of the many challenges confronting humanity in the 21st century, few are likely to prove as important—or as daunting—as global climate change. Central to the dilemma is the acrimonious debate surrounding it, particularly over the degree to which man contributes to the phenomenon. Most climatologists contend that the planet is warming in whole or in part because of fossil-fuel consumption—the burning of coal, oil, and gas for energy—which releases carbon dioxide, a heat-trapping gas, into the atmosphere, thus exacerbating the so-called greenhouse effect. Skeptics refute this case in one of two ways: Some argue that the planet is not warming at all, that faulty instruments—or highly speculative computer models—have led scientists to erroneous conclusions; others maintain that while the Earth may indeed be getting hotter, it is due to natural variations rather than man's impact on the environment. At the same time, there are those who acknowledge a human role in climate change, but believe that the economic cost of counteracting it—by promoting conservation or developing alternative energy sources, for example—would ultimately exact a higher toll than the warming itself.

This controversy is so compelling because it illuminates themes and conflicts that are central to the human condition. Chief among them is man's existential fear of the end of the world. Much like the threat of nuclear holocaust during the Cold War, climate change has aroused in many people profound concerns about the future of humanity and of the planet as a whole. The motion picture *The Day After Tomorrow*, though seemingly far-fetched, brought these insecurities into stark relief, as it depicted global warming ushering in a new ice age, with catastrophic results for mankind. Embedded in the climate-change debate is also a hint of pastoral longing, of a fundamental discomfort with the fruits of the industrial revolution and a desire to return the Earth to its natural state. Global warming skeptics, on the other hand, are often perceived as shortsighted and possessed of an irresponsible form of denial, crass human greed, and/or an almost willful desire to see the world ruined. Amid these broad themes, there is likewise a great deal of misinformation and specious reasoning. For example, laymen often misunderstand what the term "global warming" means, that it does not imply that temperatures are rising everywhere, but only that the worldwide average is increasing. This misconception leads nonscientists to conclude from a particularly cold month or after an especially intense blizzard that global warming does not exist. Similarly, a harsh hurricane season prompts many to jump to the conclusion that climate change is to blame, with or without any concrete evidence.

All of these issues, however, ultimately obscure the underlying scientific scholarship—the hard evidence—that suggests global warming is more than a myth. Indeed, the debate over whether climate change exists is essentially political and philosophical in nature—that is, the scientific debate has largely been settled: The broad consensus is that the Earth is warming and that human fossil-fuel consumption is one of the prime culprits. This consensus may, of course, turn out to be wrong, but until a new and better understanding emerges, it remains difficult to ignore.

This book explores the science undergirding global warming, as well as the arguments of climate-change skeptics, and charts how the phenomenon has already affected the planet and how it is likely to do so in the future. In addition, the question of what, if anything, man can do to offset or neutralize global warming is examined. The articles found in the first chapter, "Science and Skepticism: Climate Change and the Greenhouse Effect," catalog the scientific principles of global warming, as well as the emotional debate and the accompanying media coverage that many perceive as obscuring that science. Through an exploration of melting polar ice, rising regional temperatures, and other environmental realities, the second chapter, "Vanishing Glaciers, Rising Tides: Global Warming's Toll Thus Far," analyzes the effect climate change has already had in several areas of the world. What the years ahead are likely to bring should global warming continue unabated is the topic of the third section, "Armageddon Approaching? What the Future Holds." The final chapter, "What Can Be Done? Kyoto and Beyond," details a number of proposals aimed at slowing or halting the pace of global climate change; particular articles explore the feasibility of using alternative energy sources, like nuclear power or hydrogen cells, in place of fossil fuels, while other entries discuss how conservation agreements like the Kyoto Protocol may mitigate the impact of global warming.

The appendix contains the text of the historic Kyoto Protocol. Initially drafted in 1997, the agreement went into effect in 2005. Though set to expire in 2012, as the first treaty of its kind, the protocol will likely be the model upon which future climate-change agreements are based.

In conclusion, I would like to thank the many writers and publishers who have kindly granted permission to reprint their works here. In addition, I would like to acknowledge the many friends and colleagues at the H. W. Wilson Company who have contributed their talent and energy to this endeavor, particularly Jennifer Curry, Lynn Messina, Michael Schulze, and Sandra Watson.

Paul McCaffrey
February 2006

I. Science and Skepticism: Climate Change and the Greenhouse Effect

Editor's Introduction

The selections in this chapter help explain the scientific basis of global climate change and provide an overview of the contentious debate that frequently overshadows it. The first article, "Heat: How We Got Here," by Ian Sample, describes the history of climate-change science, focusing on the work of such pioneering researchers as Joseph Fourier and Svante Arrhenius, who, in the 19th century, developed the underlying theorems that are believed to govern the process of global warming. Fourier discovered that various gases, carbon dioxide among them, "shrouded the planet like a bell jar, transparent to sunlight, but absorbing to infrared rays." This means that "the atmosphere is heated from above and below: first by sunlight as it shines through and second by the infrared the Earth emits as it cools overnight." Without this heat-regulating process—dubbed the greenhouse effect—the planet would be uninhabitable, subject to scorching sun during the day and sub-zero temperatures at night. Arrhenius built on Fourier's scholarship as he sought to uncover what natural mechanisms caused ice ages, ultimately revolutionizing our understanding of greenhouse gases, particularly the degree to which varying concentrations of water vapor and carbon dioxide impact the climate. His calculations established that increased carbon dioxide levels led to a corresponding increase in temperatures and, consequently, that man's burning of coal and other fossil fuels could exacerbate the greenhouse effect.

Roger Di Silvestro lays out the considerable empirical evidence supporting the existence of anthropogenic—human-influenced—global warming in the second piece, "The Proof Is in the Science." He notes that the average global temperature has increased by 1.4 degrees Fahrenheit since 1750, leading to rising ocean levels and shrinking glaciers, and that the National Academy of Sciences and numerous other eminent scholarly organizations have concluded that human behavior is in large measure responsible for these changes. On the other hand, in the subsequent entry, "Strange Science," Thomas Sieger Derr, a professor of religion and ethics at Smith College, offers a multifaceted critique of this evidence, as well as the motives of those he terms "purveyors of climate disaster theories." He states that the proof of rising temperatures is not necessarily accurate, and even if it is, the changing climate still cannot be attributed to man. Moreover, Derr contends, rising temperatures could turn out to be a positive development, bringing about an increase in the planet's agricultural output. He also assails the measures proposed to counteract global climate change, particularly the Kyoto Protocol.

Jules and Maxwell Boykoff discuss the results of a study analyzing how the mainstream media covers the global-warming controversy in the next article, "Journalistic Balance as Global Warming Bias." According to the study, press reports frequently leave the mistaken impression that the scientific community is closely divided as to whether anthropogenic climate change exists, when there is actually a consensus that human beings do indeed contribute to global warming.

In "The New Extreme Sport," William Burroughs presents a more complicated assessment of the global-warming threat. He acknowledges climate change as a reality, but he argues that "we must not exaggerate current weather extremes."

Global-warming skeptics frequently contend that the climate-change models put forward by scientists anticipated a larger disparity between atmospheric and ground temperatures than has been consistently found. However, researchers subsequently discovered that the discrepancy arose due to faulty data gathered from satellites and weather balloons, as a writer for Greenwire reports in this chapter's final entry, "Climate Change: Errors in Temperature Data Mask Evidence of Warming—Studies." When scholars accounted for this discrepancy, they found that global warming was actually occurring at a faster rate than the earlier models had anticipated. Nevertheless, this revelation has neither silenced global-warming skeptics nor engendered a more vigorous response from public policy makers.

Heat

How We Got Here

By Ian Sample
The Guardian (London), June 30, 2005

Behind the treelined embankment that borders the campus of Stockholm University lies building 92E, a red brick villa as big as a fire station, its back turned to Roslagsvagen, the main artery linking the capital city with Norrtalje 70 km away.

What few markings there are on the building suggest nothing of its history. A sign above the entrance identifies it as Cafe Bojan, a student canteen, and a few shirtless students on a bench in the morning sun recall it as nothing more.

At the end of the 19th century, building 92E was the home and laboratory of Svante Arrhenius, a chemist who became Sweden's first Nobel prizewinner. He was destined to have a bigger impact than he could have imagined, far beyond his mainstream work. Unwittingly, he uncovered secrets of the Earth's atmosphere and in doing so triggered research into what many see as the biggest threat to modern humans. He is arguably the father of climate change science.

That title would be a surprise, even to him. The son of a land surveyor, Arrhenius thrived at school, showing a particular aptitude for arithmetic, but his diversity of thought and penchant for maverick theories dealt him a hefty blow at university. His PhD research, which he began at Uppsala University to the north of Stockholm, focused on the conductivity of electrolytes, but the ideas he put forward in his thesis baffled his professors and he was awarded the lowest possible pass grade. At once, any hopes of staying on at Uppsala were destroyed, and Arrhenius embarked on a tour of European laboratories before landing a job in Stockholm several years later.

Arrhenius became interested in a debate occupying the scientific community, namely the cause of the ice ages. Could it be, he wondered, that vast swings in the levels of atmospheric CO_2, lasting tens of millions of years, were the trigger?

The link between CO_2 and the Earth's temperature had been made years beforehand. It was the French scientist Joseph Fourier who first realised that certain atmospheric gases shrouded the planet like a bell jar, transparent to sunlight, but absorbing to infrared rays. It means the atmosphere is heated from above and below: first, by sunlight as it shines through and second by the infrared the Earth emits as it cools overnight.

Arrhenius set himself the task of working out just how much water and CO_2 in the atmosphere warmed the planet. From others' work, he knew that CO_2 was only part of the process. While CO_2 and other gases trapped infrared radiation and so heated the atmosphere, warmer air holds more water vapour, itself the most potent contributor to the greenhouse effect. So, if atmospheric CO_2 levels increased, water vapour would ensure the warming effect was seriously magnified.

What followed was a year doing what Arrhenius described as "tedious calculations." His starting point was a set of readings taken by US astronomer Samuel Langley, who had tried to work out how much heat the Earth received from the full moon. Arrhenius used the data with figures of global temperatures to work out how much of the incoming radiation was absorbed by CO_2 and water vapour, and so heated the atmosphere.

Between 10,000 and 100,000 calculations later, Arrhenius had some rough, but useful, results that he published in 1896. If CO_2 levels halved, he concluded, the Earth's surface temperature would fall by 4–5°C. There was a flipside to his calculations: doubling CO_2 levels would trigger a rise of about 5–6°C.

Beyond the argument over ice ages it wasn't lost on Arrhenius that human activity, in the form of widespread burning of coal, was pumping atmospheric CO_2 above the natural levels that help make the Earth habitable. Almost as a passing comment, he estimated that coal burning would drive a steady rise in CO_2 levels of about 50% in 3,000 years, a prospect he found entirely rosey. At a lecture that same year, he declared: "We would then have some right to indulge in the pleasant belief that our descendants, albeit after many generations, might live under a milder sky and in less barren surroundings than is our lot at present."

As the first to put hard figures on the greenhouse effect, it's unsurprising Arrhenius's estimates weren't spot on. He thought it would take millenia to see a 50% rise in CO_2—but modern measurements show a 30% rise during the 20th century alone. He thought a doubling of CO_2 would raise temperatures by 5–6°C. Scientists now say 2–3°C is more likely.

Over the next decades, his work was criticised, backed up and criticised again. Many disregarded his conclusions, pointing to his simplification of the climate and how he failed to account for changes in cloud cover and humidity. The oceans would absorb any extra CO_2 pumped into the atmosphere, and any remainder would be absorbed by plant life, leading to a more lush landscape, sceptics argued.

In 1938, nine years after Arrhenius had died a Nobel prizewinner for his work on ionic solutions, English engineer Guy Callendar gave the greenhouse theory a boost. An expert on steam technology, he took up meteorology as a sideline and became interested in suggestions of a warming trend. Callendar pieced together temperature measurements from the 19th century onwards and saw an appreciable rise. He went on to check CO_2 over the same period and discovered levels had increased about 10% in 100 years. The warming was probably due to the higher levels of CO_2.

The existence of an increasing greenhouse effect was hotly debated until the postwar funding of the 1950s kicked in and researchers began to get firm data. In 1956, physicist Gilbert Plass confirmed adding CO_2 to the atmosphere would increase infrared radiation absorbed, adding that industrialisation would raise the Earth's temperature by just over 1°C per century. By the end of the 1950s, Plass and other scientists in the US started warning government officials that greenhouse warming might become a serious issue in the future.

Few saw the greenhouse effect and the warming it would bring as being a problem.

Unwittingly, the US especially had already started monitoring what many believed were the direct effects of a warming world. Submarines operating in the Arctic Circle took accurate readings of the thickness of the ice sheets above them. When the Pentagon released the data nearly 40 years later, it revealed a startling melting of the ice, on average a 40% thinning of 1.3m since 1953.

In the 1960s, researchers at Scripps Institution of Oceanography in San Diego took on the testing challenge of taking a vast number of measurements of atmospheric CO_2. The aim was to establish a baseline level with which future readings in a decade or so could be compared.

Charles Keeling spent two years taking measurements in Antarctica and above the Mauna Loa volcano in Hawaii but reported that even in this short period, CO_2 levels had risen. He concluded that the oceans weren't absorbing greenhouse gases being pumped out by industry. Instead, emissions were driving levels of CO_2 higher. "It was a seminal discovery. For the first time, scientists knew that the oceans weren't going to absorb all this carbon dioxide," says Mike Hulme at the Tyndall Centre for climate change research at the University of East Anglia.

Still, few saw the greenhouse effect and the warming it would bring as being a problem. At the time, computer models were suggesting modest increases, perhaps 2°C in hundreds of years.

By the 1980s, climate change had become a megascience, attracting scientists from diverse fields, each attacking the problem from a different angle. One technique was especially useful. Deep cores of ice cut from Greenland and elsewhere held pockets of air dating back hundreds of thousands of years. By analysing the trapped air, scientists worked out CO_2 levels in the atmosphere during past ice ages. In 1987, a core cut from central Antarctica showed that in the previous 400,000 years, CO_2 had dropped to 180 parts per million (ppm) during the most extreme glacial periods and climbed as high as 280ppm in warmer times, but not once had been higher. In the outside air, CO_2 was measured at 350ppm, unprecedented for nearly half a million years.

To mainstream scientists, evidence that warming was down to human activity was becoming too big to ignore. While scientists uncovered evidence for the greenhouse effect and warming it was producing, others pointed to different processes impacting on global climate. Volcanos, for example, blast millions of tonnes of sulphur dioxide into the atmosphere that form aerosol particles which reflect sunlight back into space. The 1991 eruption of Mount Pinatubo in the Phillipines sent about 20m tonnes of the gas into the atmosphere, leading to a global cooling of around 0.5°C a year later. Scientists now believe that the warming experienced in the early 20th century can largely be explained by the lack of volcanic activity.

Variations in the sun's intensity have also been fingered as a driver of climate change. According to Joanna Haigh at Imperial College London, about a third of the warming since 1850 can be explained by solar activity. The identification of disparate contributors to warming has been seized upon by a minority who claim global warming is driven far more by nature than human activity, and the ensuing controversy is still not settled.

By 1988, the United Nations had established the Intergovernmental Panel on Climate Change to review relevant research. The panel's latest estimate points to a warming of 1.4–5.8°C by 2100, depending on what strategies, if any, are adopted to curb emissions. The 20th century saw a rise in temperature of 0.6°C, about half of which occurred since 1970.

Arguably the most concerted effort to cut global emissions has been triggered by the Kyoto Protocol. Since ratification began in 1997, more than 100 countries have adopted the protocol, which for the first time committed them to cutting emissions of six greenhouse gases.

Now, barely a week goes by without a major study on climate change. A flurry of papers started the year with warnings that the Gulf Stream would grind to a halt, ski resorts would move to higher altitudes and Antarctic glaciers were melting fast. More than 100 years after Arrhenius set out to discover why the world fell into periodic ice ages, the scientist has become a pillar of the megascience that is global warming research.

Back in Stockholm's meteorology department, Erland Kallen is musing about progress since Arrhenius first set about his calculations. "Even when I came to this field 20 years ago, I was very sceptical about global warming. There were too many uncertainties I just couldn't see how anyone could say anything sensible about it. Now, I struggle to see what other explanation there could be."

The Proof Is in the Science

By Roger Di Silvestro
NATIONAL WILDLIFE, April/May 2005

The twentieth century was the warmest of the past 1,000 years, and 19 of the 20 hottest years on record occurred after 1980. Most of this warming resulted from human activities, not nature, according to the United Nations' Intergovernmental Panel on Climate Change. Composed of 1,500 climatologists, the panel was created in 1988 by the World Meteorological Organization and the United Nations Environmental Program to evaluate climate science as a basis for setting policy. Other groups that agree with the panel's conclusion include the National Academy of Sciences, the American Meteorological Society, the American Geophysical Union and the American Association for the Advancement of Science.

The chief culprit in global warming is increased atmospheric carbon dioxide from industries and motor vehicles—at 372 parts per million, atmospheric carbon dioxide is now at the highest concentration in at least 420,000 years, as indicated by studies of gases trapped in ancient ice. This rise in density turns the atmosphere into an increasingly heavy blanket, allowing it to hold in more of the sun's heat rather than letting it radiate back into space.

So far, the global average temperature has risen 1.4 degrees F since 1750, a significant amount in terms of the world's overall average. In January, a study involving 95,000 participants from 150 countries—the world's largest climate-prediction experiment ever—concluded that greenhouse gases could raise global temperatures as much as 20 degrees F by 2100. The result: major droughts, sea-level rise and crop failures.

This prediction should raise grave concerns among policymakers, because other predictions that scientists have made in recent years about the ecological effects that rising temperatures would produce are coming true, confirming that global warming is here. Among the forecasts: warming will take place most rapidly and intensively at the poles, glaciers and ice sheets will melt, sea level will rise, precipitation patterns will change, storms and floods will become more frequent and severe, and some plants and animals will shift their ranges northward or up mountainsides to escape rising heat.

The following compendium of data, based on information from peer-reviewed scientific sources, cites forecasts that have already evolved into solid evidence for the advent of global warming.

Polar Change

The rate of warming in the Arctic was eight times faster during the past 20 years than during the previous 100 years and is occurring at nearly twice the rate of the rest of the planet. Average winter temperatures in Alaska and western Canada have risen by as much as 7 degrees F during the past 60 years.

Shrubs have been moving into the tundra, where cold temperatures historically have kept them out. Because shrubs absorb more solar heat than does tundra, they may compound the effects of global warming there.

Average temperatures in the Antarctic have increased by as much as 4.5 degrees F since the 1940s, among the fastest rates of change in the world.

Glaciers and Ice Sheets

Antarctica's 1,200-acre Larsen B ice shelf—more than 700 feet thick—collapsed in March 2002, the third large shelf in the area to do so since 1995. The calving of icebergs from ice shelves is a normal event, but complete disintegration of shelves that scientists believe may have been as much as 12,000 years old is not.

The extent of Arctic sea ice in September 2004 was more than 13 percent below average, yielding the most shrunken sea ice of the past half century. The sea ice now is melting 20 percent faster than it did two decades ago.

The lowest elevation for freezing among mid-latitude mountains, such as the U.S. Rockies and the European Alps, has shifted upward by almost 500 feet since 1970. Some 80 percent of the snowcap of Kenya's Mount Kilimanjaro has disappeared, and the 150 glaciers that studded Glacier National Park in Montana in 1910 have been reduced to fewer than 30.

Weather Events

Annual precipitation in southern New England has increased by more than 25 percent during the past century, while snowfall in northern New England has decreased by 15 percent since 1953. Snow lies on the ground in New England for seven days less per year than it did 50 years ago.

Severe droughts today affect 30 percent of the Earth's surface, compared to 10 to 15 percent 35 years ago, a change that climate scientists blame in large part on rising temperatures.

Snowfall in Australia has declined by 30 percent during the past 40 years.

Lakes in Pennsylvania freeze on average about 10 days later than they did 50 years ago and thaw about 9 days earlier.

Sea Level

Sea temperatures have risen up to 2 degrees F during the past 20 years, although the link to global warming has yet to be determined. The mean global sea level has risen as much as 7.8 inches during the past century.

Shifting Species

In Russia's Ural Mountains, the tree line has moved as much as 500 feet higher since the start of the 20th century. In Canada's Banff National Park, spruces have shifted upward by 150 to 180 feet since 1990.

Data collected for 100 flowering species in the Washington, D.C., area reveal that 89 species now are blossoming an average of 4.5 days earlier than they did in 1970. Only 11 are flowering later. In Edmonton, Alberta, a similar study found that overall spring flowering in the area occurs eight days earlier than it did 60 years ago.

The vegetative growing season of trees, shrubs and herbs in Europe has increased by 11 days since 1960.

Oak trees in England are leafing two weeks earlier than they did 40 years ago. Great white sharks and Portuguese man o'war jellyfish are moving into waters off Devon and Cornwall, previously too cold for these species.

A study of some 1,700 species completed in 2002 found that some birds and butterflies have shifted their ranges northward by 4 miles yearly since the 1960s.

An analysis of 74,000 nesting records from 65 bird species in the United Kingdom found that between 1971 and 1995, 20 of the species were laying their first eggs an average of nine days earlier.

In just 35 years, the sachem skipper butterfly has expanded its range 420 miles northward, from California into Washington. During 1998, the warmest year on record, the range expanded 75 miles.

Other U.S. species that have expanded northward include the red fox, rufous hummingbird, two subtropical dragonflies in Florida and a variety of marine species found off Monterey, California, that are following warming seas away from the equator.

"No single bit of scientific evidence makes a convincing argument that global warming is having an impact on wildlife and plants," says Doug Inkley, NWF senior science advisor, "but the cumulative evidence cannot be ignored. The question is no longer 'Is global warming happening?' The question is, what are we going to do about it?"

Strange Science

By Thomas Sieger Derr
First Things, November 2004

Global warming has achieved the status of a major threat. It inspires nightmares of a troubled future and propels apocalyptic dramas such as the summer 2004 movie *The Day After Tomorrow.* Even were the Kyoto treaty to be fully implemented, it wouldn't make a dent in the warming trend, which seems to be inexorable. Doom is upon us.

Except that maybe it isn't. You might not know it from ordinary media accounts, which report the judgments of alarmists as "settled science," but there is a skeptical side to the argument. Scientists familiar with the issues involved have written critically about the theory of global warming. The puzzle is why these commentators, well-credentialed and experienced, have been swept aside to produce a false "consensus." What is it that produces widespread agreement among both "experts" and the general public on a hypothesis which is quite likely wrong?

The consensus holds that we are experiencing unprecedented global warming and that human activity is the main culprit. The past century, we are told, has been the hottest on record, with temperatures steadily rising during the last decades. Since human population and industrial activity have risen at the same time, it stands to reason that human activity is, one way or another, the cause of this observed warming. Anything wrong with this reasoning?

Quite a lot, as it turns out. The phrase "on record" doesn't mean very much, since most records date from the latter part of the nineteenth century. Without accurate records there are still ways of discovering the temperatures of past centuries, and these methods do not confirm the theory of a steady rise. Reading tree rings helps (the rings are further apart when the temperature is warmer and the trees grow faster). Core samples from drilling in ice fields can yield even older data. Some historical reconstruction can help, too—for example, we know that the Norsemen settled Greenland (and named it "green") a millennium ago and grew crops there, in land which is today quite inhospitable to settlement, let alone to agriculture. Other evidence comes from coral growth, isotope data from sea floor sediment, and insects, all of which point to a very warm climate in medieval times. Abundant testimony tells us that the European climate then cooled dramatically from the thirteenth century until the eighteenth, when it began its slow rewarming.

In sum, what we learn from multiple sources is that the earth (and not just Europe) was warmer in the tenth century than it is now, that it cooled dramatically in the middle of our second millennium (this has been called the "little ice age"), and then began warming again. Temperatures were higher in medieval times (from about 800 to 1300) than they are now, and the twentieth century represented a recovery from the little ice age to something like normal. The false perception that the recent warming trend is out of the ordinary is heightened by its being measured from an extraordinarily cold starting point, without taking into account the earlier balmy medieval period, sometimes called the Medieval Climate Optimum. Data such as fossilized sea shells indicate that similar natural climate swings occurred in prehistoric times, well before the appearance of the human race.

Even the period for which we have records can be misread. While the average global surface temperature increased by about 0.5 degrees Celsius during the twentieth century, the major part of that warming occurred in the early part of the century, before the rapid rise in human population and before the consequent rise in emissions of polluting substances into the atmosphere. There was actually a noticeable cooling period after World War II, and this climate trend produced a rather different sort of alarmism—some predicted the return of an ice age. In 1974 the National Science Board, observing a thirty-year-long decline in world temperature, predicted the end of temperate times and the dawning of the next glacial age. Meteorologists, *Newsweek* reported, were "almost unanimous in the view that the trend will reduce agricultural productivity for the rest of the century." But they were wrong, as we now know (another caution about supposedly "unanimous" scientific opinion), and after 1975 we began to experience our current warming trend. Notice that these fluctuations, over the centuries and within them, do not correlate with human numbers or activity. They are evidently caused by something else.

What, then, is the cause of the current warming trend? As everyone has heard, the emission of so-called "greenhouse gasses," mostly carbon dioxide from burning fossil fuels, is supposed to be the major culprit in global warming. This is the anthropogenic hypothesis, according to which humans have caused the trouble. But such emissions correlate with human numbers and industrial development, so they could not have been the cause of warming centuries ago, nor of the nineteenth-century rewarming trend which began with a much smaller human population and before the industrial revolution. Nor is there a very good correlation between atmospheric carbon dioxide levels and past climate changes. Thus, to many scientists, the evidence that greenhouse gasses produced by humans are causing any significant warming is sketchy.

The likeliest cause of current climate trends seems to be solar activity, perhaps in combination with galactic cosmic rays caused by supernovas, especially because there is some good observable correlation between solar magnetism output and terrestrial climate change. But that kind of change is not predictable within any usable time frame, not yet

> *Global warming is not necessarily a bad thing.*

anyway, and, of course, it is entirely beyond any human influence. The conclusion, then, is that the climate will change naturally; aside from altering obviously foolish behavior, such as releasing dangerous pollutants into our air and water, we can and should do little more than adapt to these natural changes, as all life has always done.

That is not a counsel of despair, however, for global warming is not necessarily a bad thing. Increasing warmth and higher levels of carbon dioxide help plants to grow (carbon dioxide is not a pollutant), and, indeed, mapping by satellite shows that the earth has become about six percent greener overall in the past two decades, with forests expanding into arid regions (though the effect is uneven). The Amazon rain forest was the biggest gainer, despite the much-advertised deforestation caused by human cutting along its edges. Certainly climate change does not help every region equally and will probably harm some. That has always been true. But there are careful studies that predict overall benefit to the earth with increasing warmth: fewer storms (not more), more rain, better crop yields over larger areas, and longer growing seasons, milder winters, and decreasing heating costs in colder latitudes. The predictable change, though measurable, will not be catastrophic at all—maybe 1 degree Celsius during the twenty-first century. The news is certainly not all bad, and may on balance be rather good.

There is much more, in more detail, to the argument of those scientists who are skeptical about the threat of global warming. On the whole, their case is, I think, quite persuasive. The question, then, is why so few people believe it.

Part of the answer is that bad news is good news—for the news media. The media report arresting and frightening items, for that is what draws listeners, viewers, and readers. The purveyors of climate disaster theories have exploited this journalistic habit quite brilliantly, releasing steadily more frightening scenarios without much significant data to back them up. Consider the unguarded admission of Steven Schneider of Stanford, a leading proponent of the global warming theory. In a now notorious comment, printed in *Discover* in 1989 and, surely to his discomfort, often cited by his opponents, Schneider admitted:

> To capture the public imagination, we have to offer up scary

scenarios, make simplified dramatic statements, and make little mention of any doubts we may have. Each of us has to decide what the right balance is between being effective and being honest.

This sort of willingness to place the cause above the truth has exasperated Richard Lindzen, Sloan Professor of Meteorology at MIT, who is one of the authors of the science sections of the report of the International Panel on Climate Change (IPCC), the body responsible for an increasing crescendo of dire warnings. In testimony before the U.S. Senate's Environment and Public Works Committee, he called the IPCC's Summary for Policymakers, which loudly sounds the warming alarm, "very much a child's exercise of what might possibly happen . . . [which] conjures up some scary scenarios for which there is no evidence."

This brings us to the second part of the answer, which concerns the political and economic consequences of the policy argument. The IPCC is a UN body and reflects UN politics, which are consistently favorable to developing countries, the majority of its members. Those politics are very supportive of the Kyoto treaty, which not

---

The Kyoto treaty would not make a measurable difference in the climate.

---

only exempts the developing countries from emissions standards but also requires compensatory treatment from the wealthier nations for any economic restraints that new climate management policies may impose on these developing countries. Were Kyoto to be implemented as written, the developing countries would gain lots of money and free technology. One need not be a cynic to grasp that a UN body will do obeisance to these political realities wherever possible.

The Kyoto treaty would not make a measurable difference in the climate—by 2050, a temperature reduction of maybe two-hundredths of a degree Celsius, or at most six-hundredths of a degree—but the sacrifices it would impose on the United States would be quite large. It would require us to reduce our projected 2012 energy use by 25 percent, a catastrophic economic hit. Small wonder that the Senate in 1997 passed a bipartisan resolution, the Byrd-Hagel anti-Kyoto resolution, by 95-0 (a fact rarely recalled by those who claim that America's refusal to sign on to the treaty was the result of the Bush administration's thralldom to corporate interests).

Most of the European countries that have ratified Kyoto are falling behind already on targets, despite having stagnant economies and falling populations. It is highly unlikely they will meet the goals they have signed on for, and they know it. Neither will Japan, for that matter. The European Union has committed itself to an eight percent reduction in energy use (from 1990 levels) by 2012, but the

European Environment Agency admits that current trends project only a 4.7 percent reduction. When Kyoto signers lecture nonsigners for not doing enough for the environment, they invite the charge of hypocrisy. There is also the obvious fact that adherence to the treaty will hurt the U.S. economy much more than the European, which suggests that old-fashioned economic competitiveness is in the mix of motives at play here. The absurdity of the treaty becomes obvious when we recognize that it does not impose emissions requirements on developing countries, including economic giants such as China, India, and Brazil. (China will become the world's biggest source of carbon dioxide emissions in just a few years.)

A third reason why global warming fears seem to be carrying the day goes beyond these political interests; it involves intellectual pride. Academics are a touchy tribe (I'm one of them); they do not take it kindly when their theories, often the result of hard work, are contradicted. And sure enough, the struggle for the truth in this matter is anything but polite. It is intellectual warfare, entangled with politics, reputations, and ideology; and most of the anger comes from the side of the alarmists. People lose their tempers and hurl insults—"junk science," "willful ignorance," "diatribe," "arrogant," "stupid," "incompetent," "bias," "bad faith," "deplorable misinformation," and more. Consider the fiercely hateful reaction to Bjorn Lomborg's 2001 book, *The Skeptical Environmentalist*. He challenged the entrenched and politically powerful orthodoxy and did so with maddeningly thorough data. His critics, unable to refute his statistics, seem to have been enraged by their own weakness—a familiar phenomenon, after all. Or perhaps, with their reputations and their fund-raising ability tied to the disaster scenarios, they felt their livelihoods threatened. In any case, the shrillness of their voices has helped to drown out the skeptics.

Finally, there is a fourth cause: a somewhat murky antipathy to modern technological civilization as the destroyer of a purer, cleaner, more "natural" life, a life where virtue dwelt before the great degeneration set in. The global warming campaign is the leading edge of an environmentalism which goes far beyond mere pollution control and indicts the global economy for its machines, its agribusiness, its massive movements of goods, and above all its growing population. Picking apart this argument to show the weakness of its pieces does not go to the heart of the fear and loathing that motivate it. The revulsion shows in the prescriptions advanced by the global warming alarmists: roll back emissions to earlier levels; reduce production and consumption of goods; lower birth rates. Our material ease and the freedoms it has spawned are dangerous illusions, bargains with the devil, and now comes the reckoning. A major apocalypse looms, either to destroy or, paradoxically, to save us—if we come to our senses in the nick of time.

It is clear, then, given the deep roots of the scare, that it is likely to be pretty durable. It has the added advantage of not being readily falsifiable in our lifetimes; only future humans, who will have the perspective of centuries, will know for certain whether the current warming trend is abnormal. In the meantime, the sanest course for us would be to gain what limited perspective we can (remembering the global cooling alarm of a generation ago) and to proceed cautiously. We are going through a scarce with many causes, and we need to step back from it, take a long second look at the scientific evidence, and not do anything rash. Though the alarmists claim otherwise, the science concerning global warming is certainly not settled. It is probable that the case for anthropogenic warming will not hold up, and that the earth is behaving as it has for millennia, with natural climate swings that have little to do with human activity.

Journalistic Balance as Global Warming Bias

BY JULES BOYKOFF AND MAXWELL BOYKOFF
EXTRA!, NOVEMBER/DECEMBER 2004

A new study has found that when it comes to U.S. media coverage of global warming, superficial balance—telling "both" sides of the story—can actually be a form of informational bias. Despite the consistent assertions of the United Nations–sponsored Intergovernmental Panel on Climate Change (IPCC) that human activities have had a "discernible" influence on the global climate and that global warming is a serious problem that must be addressed immediately, "he said/she said" reporting has allowed a small group of global warming skeptics to have their views greatly amplified.

The current best climate research predicts that the Earth's temperature could rise by as much as 10.4°F by 2100. Studies show that this temperature increase could contribute to a sea-level rise of up to 35 inches by 2100—threatening to flood tens of millions of inhabitants of coastal communities. Warming on this scale would extend the range and activity of pests and diseases, and force land and marine life to migrate northward, thereby endangering ecosystems, reproductive habits and biodiversity.

Moreover, climate forecasts include more and higher-intensity rainfall in some regions, leading to greater flood and landslide damage. In other regions, forecasts call for increased droughts, resulting in smaller crop yields, more forest fires and diminished water resources. These climate shifts threaten the lives and livelihoods of people around the globe, with a greater impact on the most vulnerable.

These gloomy findings and dire predictions are not the offerings of a gaggle of fringe scientists with an addiction to the film Apocalypse Now. Rather, these forecasts are put forth by the IPCC, the largest, most reputable peer-reviewed body of climate-change scientists in history. Formed by the United Nations in 1990 and composed of the top scientists from around the globe, the IPCC employs a decision-by-consensus approach. In fact, D. James Baker, administrator of the U.S. National Oceanic and Atmospheric Administration and undersecretary for oceans and atmosphere at the Department of Commerce under the Clinton administration, has said about human contributions to global

"Journalistic Balance as Global Warming Bias" by Jules Boykoff and Maxwell Boykoff from *Extra! The Magazine of Fairness & Accuracy in Reporting*, November/December 2004.

warming (*Washington Post*, 11/12/97) that "there's no better scientific consensus on this on any issue I know—except maybe Newton's second law of dynamics."

The Idea of Balance

In 1996, the Society of Professional Journalists removed the term "objectivity" from its ethics code (*Columbia Journalism Review*, 7–8/03). This reflects the fact that many contemporary journalists find the concept to be an unrealistic description of what journalists aspire to, preferring instead words like "fairness," "balance," "accuracy," "comprehensiveness" and "truth." In terms of viewpoints presented, journalists are taught to abide by the norm of balance: identifying the most dominant, widespread positions and then telling "both" sides of the story.

According to media scholar Robert Entman, "Balance aims for neutrality. It requires that reporters present the views of legitimate spokespersons of the conflicting sides in any significant dispute, and provide both sides with roughly equal attention."

Balanced coverage does not, however, always mean accurate coverage. In terms of the global warming story, "balance" may allow skeptics—many of them funded by carbon-based industry interests—to be frequently consulted and quoted in news reports on climate change. Ross Gelbspan, drawing from his 31-year career as a reporter and editor, charges in his books *The Heat Is On* and *Boiling Point* that a failed application of the ethical standard of balanced reporting on issues of fact has contributed to inadequate U.S. press coverage of global warming:

> The professional canon of journalistic fairness requires reporters who write about a controversy to present competing points of view. When the issue is of a political or social nature, fairness—presenting the most compelling arguments of both sides with equal weight—is a fundamental check on biased reporting. But this canon causes problems when it is applied to issues of science. It seems to demand that journalists present competing points of view on a scientific question as though they had equal scientific weight, when actually they do not.

We empirically tested Gelbspan's hypothesis as we focused on the human contribution to global warming (known in science as "anthropogenic global warming"). In our study called "Balance as Bias: Global Warming and the U.S. Prestige Press"—presented at the 2002 Conference on the Human Dimensions of Global Environmental Change in Berlin and published in the July 2004 issue of the journal *Global Environmental Change*—we analyzed articles about human contributions to global warming that appeared between 1988 and 2002 in the U.S. prestige press: the *New York Times*, *Washington Post*, *Los Angeles Times* and *Wall Street Journal*.

Using the search term "global warming," we collected articles from this time period and focused on what is considered "hard news," excluding editorials, opinion columns, letters to the editor and book reviews. Approximately 41 percent of articles came from the *New York Times*, 29 percent from the *Washington Post*, 25 percent from the *Los Angeles Times*, and 5 percent from the *Wall Street Journal*.

From a total of 3,543 articles, we examined a random sample of 636 articles. Our results showed that the majority of these stories were, in fact, structured on the journalistic norm of balanced reporting, giving the impression that the scientific community was embroiled in a rip-roaring debate on whether or not humans were contributing to global warming.

More specifically, we discovered that:

- 53 percent of the articles gave roughly equal attention to the views that humans contribute to global warming and that climate change is exclusively the result of natural fluctuations.

- 35 percent emphasized the role of humans while presenting both sides of the debate, which more accurately reflects scientific thinking about global warming.

- 6 percent emphasized doubts about the claim that human-caused global warming exists, while another 6 percent only included the predominant scientific view that humans are contributing to Earth's temperature increases.

Through statistical analyses, we found that coverage significantly diverged from the IPCC consensus on human contributions to global warming from 1990 through 2002. In other words, through adherence to the norm of balance, the U.S. press systematically proliferated an informational bias.

Global Warming 101

Building on earlier climate science work by William Herschel, John Tyndall and Joseph Fourier, investigations regarding humans' role in global warming began in 1896, when Nobel Prize–winning physicist Svante Arrhenius examined contributions of carbon dioxide emissions to increases in atmospheric temperature. In the 1930s, meteorologist G. S. Callendar gathered temperature records from more than 200 weather stations around the world and attributed temperature increases to greenhouse gas emissions from industry.

In the 1950s, Gilbert Plass' research on atmospheric CO_2 and infrared radiation absorption added to a growing scientific consensus that humans contribute to global warming. In 1956, Plass announced that human activities were raising the average global temperature.

Also, beginning in 1958, Charles David Keeling began to document atmospheric carbon dioxide levels from Mauna Loa volcano in Hawaii. His findings of a dramatic increase in CO_2—referred to as the "Keeling Curve"—are considered some of the most important long-term data relating to humans' role in global warming. Additionally, 1966 and 1977 United States National Academy of Sciences reports made clear links between human activities and global warming.

NASA scientist James Hansen's 1988 testimony to the U.S. Congress marked solidified scientific concern for human-caused global warming. He said he was "99 percent certain" that warmer temperatures were caused by the burning of fossil fuels and not solely a result of natural variation and that "it is time to stop waffling so much and say that the evidence is pretty strong that the greenhouse effect is here."

Since the formation of the IPCC in 1988 by the United Nations Environment Program and the World Meteorological Organization, a steady flow of IPCC reports have continued to support the notion that humans are contributing to global warming. For example, in 1990 at the World Climate Conference in Geneva, over 700 scientists from around the world gathered to review the IPCC First Scientific Assessment Report in order to set the stage for the crafting of the 1992 United Nations Framework Convention on Climate Change (UNFCCC). After their review, they released the Scientists' Declaration, which focused on human-caused global warming, and read, "A clear scientific consensus has emerged on estimates of the range of global warming that can be expected during the 21st century. . . . Countries are urged to take immediate actions to control the risks of climate change." Another salient assertion regarding human contributions to warming manifested in the Second Scientific Assessment Report, released in 1995. The consensus statement strongly asserted that there has been "a discernible human influence" on the global climate.

Balanced to a Fault

Specific examples abound that demonstrate a contrast between "balanced reporting" in newspaper coverage and this scientific consensus on human-caused global warming. For example, an article that appeared on the front page of the *Los Angeles Times* (12/2/92) reported:

> The ability to study climatic patterns has been critical to the debate over the phenomenon called "global warming." Some scientists believe—and some ice core studies seem to indicate—that humanity's production of carbon dioxide is leading to a potentially dangerous overheating of the planet. But skeptics contend there is no evidence the warming exceeds the climate's natural variations.

Pitting what "some scientists believe" against what "skeptics contend" implies a roughly even division within the scientific community. And putting the term "global warming" in scare quotes serves to subtly cast doubt on the reality of such a phenomenon.

Another front-page *Los Angeles Times* article (2/8/93), "An Early Warning of Warming: If the 'Greenhouse Effect' Exists, the Arctic Will Be the First to Experience It," provides another example of balance as bias. After stating that "many climate experts are convinced that the world is warming up, probably because of increased atmospheric levels of 'greenhouse gases' given off by the burning of fossil fuels," the article goes on to imply a roughly even division within the scientific community:

> Such a weather log [for the Arctic] will be of tremendous help to the many scientists who are trying to find out whether the current warming trend is merely part of the natural variation in climate—or whether it is the more worrisome result of runaway fossil-fuel consumption. For those caught up in the global-warming debate, this is the threshold question.
>
> The evidence so far is inconclusive. Scientists agree that the levels of carbon dioxide in the atmosphere have increased by about 25 percent over the past century. And credible statistics support a finding that not only is the Earth warming but that the past decade was, on average, the warmest since record-keeping began in the latter part of the 19th century.
>
> But is there a clear connection between the rise in carbon dioxide concentrations and the warming temperature? That's where many competent researchers admit they are stumped. They point out that the Earth has gone through other warm spells down through the eons, none of them brought on through human deeds. Today's rising temperatures, they say, may just be another one of those natural fluctuations.

Aside from the title's insinuation that the greenhouse effect, as a scientific process, may not exist—even though this is a completely uncontroversial piece of science that explains why atmospheres tend to warm planets—the article also portrays a balanced debate on whether global warming is caused by fossil-fuel emissions.

Yet another example of this balance-as-bias phenomenon comes from a 1995 *Washington Post* article (3/28/95) that previewed the First Conference of the Parties (COP1) to the U.N. Framework Convention on Climate Change in Berlin. The article described "the lack of international consensus on the causes and hazards of global warming" before turning to the concerns of residents of the Maldive Islands, a low-lying country that could be submerged if rising tides from global warming continue. After citing the distress of the Maldive president, the article closes by saying:

> On the other hand, some skeptical meteorologists and analysts assert that global warming reflects a natural cycle of temperature fluctuation and cannot be decisively tied to human actions. "As far as we are concerned, there's no

evidence for global warming, and by the year 2000 the man-made greenhouse theory will probably be regarded as the biggest scientific gaffe of the century," Piers Corbyn, an astrophysicist at London's Weather Action forecasting organization, told the Reuters news agency.

As a final example, a *Los Angeles Times* article from 2001 (4/13/01) stated:

> The issue of climate change has been a topic of intense scientific and political debate for the past decade. Today, there is agreement that the Earth's air and oceans are warming, but disagreement over whether that warming is the result of natural cycles, such as those that regulate the planet's periodic ice ages, or caused by industrial pollutants from automobiles and smokestacks.

These articles all demonstrate that adhering to the journalistic norm of balanced reporting can, in the end, lead to biased coverage.

Dueling Scientists

As we have seen, the "dueling scientists" became a common feature of the prestige-press terrain in the United States. Late in 1990, a coherent and cohesive group emerged to challenge the claims that were made in the IPCC reports. S. Fred Singer, Don Pearlman, Richard Lindzen, Sallie Baliunas, Frederick Seitz, Robert Balling Jr., Patrick Michaels and others began to speak out vociferously against the findings of the IPCC. This group is what Jeremy Leggett's book *The Carbon War* dubbed the "Carbon Club," describing them as "the foot soldiers for the fossil-fuel industries."

Scientists from the Carbon Club consistently found their way into the news. For example, in a *Washington Post* article headlined "Primary Ingredient of Acid Rain May Counteract Greenhouse Effect" (9/17/90), the skeptics were afforded prominent billing. Discussing the relative role of sulfur dioxide, the article stated:

> If the role of sulfur cooling proves to be large, and this is still far from certain, some researchers say it could be necessary to continue burning fossil fuels in order to produce sulfur dioxide to fight the carbon dioxide–driven warming. "I would not be surprised if somebody suggested concentrating fossil fuel power plants on the eastern margins of continents, which would put a lot of sulfates into the atmosphere, which would rain out over the oceans, which have a tremendous capacity to absorb acidity," [Patrick] Michaels [of the University of Virginia] said. "This plan would make sense because the prevailing winds blow from east to west."

In another article from the *New York Times* (4/22/98), another global-warming skeptic, Dr. Frederick Seitz, was portrayed as supporting a supposedly scientific study pushing the idea that carbon dioxide emissions were not a threat to the climate, but rather "a wonderful and unexpected gift from the Industrial Revolution."

These global warming skeptics deflect attention away from the IPCC's consensus on the human contributions to global warming, thereby providing space for politicians to call for "more research" before tinkering with the status-quo consumption of fossil fuels. Through "balanced" coverage, the mass media have misrepresented the scientific consensus of humans' contribution to global warming as highly divisive, what the *Washington Post* (10/31/92) once referred to as "the usual fickleness of science." Such coverage has served as a veritable oxygen supply for skeptics in both the scientific and political realms.

Through "balanced" coverage, the mass media have misrepresented the scientific consensus of humans' contribution to global warming as highly divisive.

Time for a Currency Transfer

To the surprise of many, the George W. Bush administration released a report in late August 2004 stating that carbon-dioxide emissions and other heat-trapping greenhouse gases are the most plausible explanation for global warming. Contrary to previous presidential proclamations, the report indicated that rising temperatures in North America were attributable in part to human activity and that this was having detectable effects on animal and plant life. *New York Times* environment reporter Andrew Revkin (8/26/04) dubbed this "a striking shift in the way the Bush administration has portrayed the science of climate change."

Yet despite this recent report, the Bush administration did not flinch in its stance on the issue of global warming. It continued to spurn the Kyoto Protocol, oppose actions to reduce greenhouse gas emissions from automobiles and emphasize uncertainties in the underlying climate-change science, calling for more research before taking action to curb human contributions to warming (*New York Times*, 11/13/02). In fact, John H. Marburger, Bush's science adviser, said (*Washington Post*, 8/27/04) that the most recent report has "no implications for policy." Marburger asserted, "There is no discordance between this report and the president's position on climate."

So why has the United States government—from President George H. W. Bush to Bill Clinton to George W. Bush—been so reluctant to seriously address global warming? A number of factors have contributed to this spectacular inaction: the oil and coal industries' tanker-load of annual campaign contributions to national politicians, these industries' well-connected cadre of lobbyists working Capitol Hill with aplomb, the crucial disjuncture between a scientific community that deals in a language of uncertainty and probability and a political culture that barks "If it ain't certain, it ain't real," the Bush administration's long-standing relationship with the energy industries, and so on.

But a much subtler mechanism is also at work: the journalistic norm of balanced reporting, widely considered one of the traditional pillars of good journalism. By giving equal time to opposing views, the major mainstream newspapers significantly downplayed scientific understanding of the role humans play in global warming. Certainly there is a need to represent multiple viewpoints, but when generally agreed-upon scientific findings are presented side-by-side with the viewpoints of a handful of skeptics, readers are poorly served. Meanwhile, the world dangerously warms, conservative think tanks gut the precautionary principle, and humankind—from the Carbon Club to the Boys and Girls Club—faces a dire future.

This critique is not meant as a personal attack on individual journalists. In fact, adhering to the norm of balance is a sign of professionalism, and, let's not forget, approximately 35 percent of the articles in our sample got the story correct. There are a number of journalists, such as Andrew Revkin of the *New York Times*, who are providing sound coverage of this important issue. We are more concerned with the institutional features and professional norms and practices of the mass-media system than we are with naming names of questionable journalists. Of course, these features will change when individual journalists, editors, publishers, scientists, policy makers and citizens work effectively to change them.

Clearly, the notion of balance is much more complex than it appears on the conceptual surface. Journalists have already begun the appropriate excavation of the term "objectivity." Similar archaeological work should also be carried out on "balance."

The New Extreme Sport

BY WILLIAM BURROUGHS
THE GUARDIAN (LONDON), JULY 28, 2005

Extremes in the weather have a disproportionate impact. Concern that global warming will lead to a more variable climate reinforces political resolve to act on limiting emissions. It also fuels a media obsession that every new example of freak weather is a manifestation of global warming. Is this an overreaction?

First, let me make one thing clear. Asking about extreme weather is not the same as questioning the existence of global warming. Clearly, the Earth has heated up during the 20th century, and part of this change is attributable to human activities. This is a different matter from showing that shorter-term weather fluctuations have become more frequent.

Defining more extreme weather requires careful analysis of meteorological statistics. Studies on a variety of time scales must establish whether recent events fall well outside the range of earlier experience. This calls for lengthy accurate records of, say, high and low temperatures, rainfall amounts and wind speeds. Experts from the meteorological services of Australia, the Netherlands, Britain and the US published an example of this type of analysis in 2002. It examined trends in climatic extremes during the second half of the 20th century, by distilling 3,000 continuous rainfall and temperature records from around the world.

With temperature records, the big change has been to warmer nights with a marked rise in minimum temperatures. This has led to declining nighttime frosts, lengthening of the growing season and a reduction in the extreme range between summer maxima and winter minima. Furthermore, there was no evidence of increasing severe summer heatwaves. This all adds up to a decline in the extreme nature of temperature records in the second half of the 20th century.

What about the European heatwave of 2003: widely seen as a harbinger of warmer world? Analysing Burgundy wine harvest dates since 1370 confirms that the summer of 2003 was by far the hottest summer in over 600 years in central France. But there is no clear trend to warmer summers, and the warmth of the 1990s is matched by comparable periods in the 1380s, 1420s and 1680s.

The number of days with heavy rainfall has increased, although other factors may have contributed to this rise. Conversely, there has been a decline in the number of consecutive dry days around the world; so more droughts may not be a feature of a warmer world.

Analysis of tropical storms tells the same story. While, in principle, warmer tropical oceans should produce more intense storms, in practice, there is no evidence of an upward trend. In the Atlantic, the figures are dominated by a natural variation, which produced more hurricanes between the 1940s and the 1960s, far fewer in the 1970s and 1980s, and a surge since 1995. We have longer records of winter storms across Europe, such as the storm that hit Britain in October 1987, or Lothar, which devastated France in December 1999. Contrary to popular belief, there is no evidence of a sustained rising trend in storminess.

All of this sits uneasily with the longer-term evidence of past climate change. During the past 10,000 years, the climate has been remarkably stable, but, before then, in the last ice age, it was much more extreme. This suggests that in a colder, drier global climate, the weather is far more extreme than now.

What has this to do with global warming? The answer lies in the circulation of the North Atlantic. During the last ice age, it was much less stable. The Gulf Stream switched on or off at the drop of a hat. Computer models of the climate suggest that future warming could lead to an increased influx of freshwater from the melting Greenland icesheet and heavier precipitation over Siberia that could switch the Gulf Stream off, tipping us back into a more variable climate. What is more, the Gulf Stream has slowed down in the past 10 to 15 years, although this could be no more than a natural fluctuation.

Even if warming could trigger sudden, effectively unpredictable changes, we must not exaggerate current weather extremes. Failure to take a balanced approach to these events distorts our priorities. So, next time an extreme event is cited as the result of global warming, look closely at the evidence. As the cost of reducing emissions of greenhouse gases hits our energy bills, more and more people will be asking the same question and whether it is worth the sacrifice.

Climate Change

Errors in Temperature Data Mask Evidence of Warming—Studies

GREENWIRE, AUGUST 12, 2005

Differences between surface and atmospheric temperature data are due in part to errors caused by weather balloons and satellites used to gather the measurements, according to three related analyses published yesterday in the journal *Science*.

Applying corrections for such errors to existing data shows the atmosphere has warmed more than previously thought, according to the researchers who wrote the studies.

The discrepancy between surface and atmospheric temperatures predicted by models and those observed has often been cited as evidence against global warming by its critics. But the new studies eliminate that key argument of global warming critics, said Ben Santer of Lawrence Livermore National Laboratory, an author of one of the papers.

The findings are to be featured in an upcoming report from the federal government's Climate Change Science Program, which is intended to resolve uncertainties in climate science.

In one study, researchers with climate analysis firm Remote Sensing Systems examined satellite data collected since 1979 by National Oceanic and Atmospheric Administration weather satellites. They found that the satellites' eastward drift over time threw off the timing of temperature measurements—causing many of the instruments to report nighttime temperatures as daytime ones, skewing data to indicate a false cooling trend.

"Our hats are off to [them]. They found a real source of error," said John Christy of the University of Alabama-Huntsville, who originally analyzed the NOAA satellite data (Dan Vergano, *USA Today*, Aug. 12). But Christy said he believed that the revised data still produced a warming rate too small to be a concern: "Our view hasn't changed. We still have this modest warming" (Andrew C. Revkin, *New York Times*, Aug. 12).

A second study, led by Santer, found that errors in the satellite data accounted for much of the differences in predictions in 19 existing climate models (*The Economist*, Aug. 11).

A third study by researchers at Yale University examined data collected by a worldwide network of weather balloons. Since the 1950s, the balloons have recorded temperature data twice a day, at 12 A.M. and 12 P.M. Greenwich Mean Time.

Scientists have struggled to explain why the balloon data have shown steady atmospheric temperatures, though the Earth's surface has warmed since the 1970s and climate models predict close agreement between surface and air temperatures. Part of the answer, according to Yale physicist Steven Sherwood, is error caused by the sun: Direct sun on weather balloons' sensors artificially inflates daylight temperature readings.

The problem has lessened as balloon technology has improved over the last 40 years—but to accurately analyze older readings, Sherwood and his colleagues developed a mathematical correction that subtracts the sunlight effect. With the balloon error accounted for, atmospheric data shows a 0.2 dgree Celsius temperature increase over the last 30 years, more in line with surface readings, Sherwood said.

"We've allowed the discrepancy with the more reliable surface record to be reconciled," Sherwood said. "There's no longer another data set that conflicts with the surface record. It removes a concern" about evidence for global warming often cited by "people outside of the scientific community," he said.

The corrected weather balloon data confirms human actions are major factors in Earth's warming, Sherwood said: "What's important is if you project this out to the end of the century, it comes out to something like 2 to 4 degrees Celsius [warming], and it's just going to keep going."

Unchecked Global Warming Could Negatively Affect Economy—Experts

Meanwhile, experts said this week that the U.S. economy could suffer in the long-term if the Bush administration continues to refuse placing caps on greenhouse gas emissions.

"While there are costs associated with reducing emissions, there are certainly costs associated with not doing anything," said Kevin Forbes, head of Catholic University's economics department. "It would be, in my opinion, folly not to try to do something."

"There are real economic costs associated with not taking action, including changes to water supply infrastructure, industrial capital, like pipelines, and with human health," said Janet Peace, senior research fellow at the Pew Center on Global Climate Change. "With droughts, there's a cost. With increased flooding, there's a cost [and] with increased hurricanes and tornadoes."

But White House spokeswoman Dana Perino said some of the tactics used worldwide against climate change, especially the Kyoto Protocol and other emissions cuts, have their own negative economic impacts. "We oppose policies like mandatory caps on emissions, that would achieve reductions by raising energy costs, slowing the economy, and putting Americans out of work," she said.

John Reilly of the Massachusetts Institute of Technology's Joint Program on the Science and Policy of Global Change, who studied the economic effects of several proposals, disagrees. "When we looked at implementing Kyoto . . . we estimated that would be 6/10ths or 1 percent of the economy. We thought that was costly but that's not wrecking the economy," he said.

Earth surface temperatures could be up to 10.4 degrees Fahrenheit above 1990 levels by 2100, potentially worsening storms, raising sea levels and eating away ice caps, according to National Academy of Sciences President Ralph Cicerone (Reuters, Aug. 11).

II. Vanishing Glaciers, Rising Tides: Global Warming's Toll Thus Far

Editor's Introduction

Evidence of climate change can be found throughout the world. The selections in this second chapter, "Vanishing Glaciers, Rising Tides: Global Warming's Toll Thus Far," examine much of this evidence, looking at how various natural phenomena have been impacted and how specific regions of the world are weathering the climatic shifts.

Climate change manifests itself most vividly in the polar regions. In "As the Arctic Melts, an Ancient Culture Faces Ruin," Charles Wohlforth travels toward the North Pole, to Alaska's northern coast, where he finds that temperatures have risen steadily in recent years, climbing at a much faster rate than the global average. Indeed, in some Arctic regions, winter temperatures have increased by 7 degrees Fahrenheit in just the last half century; this has led to increasingly volatile weather patterns, which have upset the area's delicate natural balance. For members of the Iñupiaq tribe, Wohlforth notes, this warming threatens to destroy their traditional way of life.

The news from Antarctica is similarly alarming, Richard A. Kerr reports in "A Bit of Icy Antarctica Is Sliding Toward the Sea." Scientists believe that climate change is destabilizing South Pole glaciers, with "potentially dangerous consequences." However, it is not just at the poles that glaciers seem to be dissipating; in "The Big Thaw," Daniel Glick finds that they are in retreat throughout the globe, with the melt water contributing to rising sea levels, which in turn pose a significant threat to densely populated coastal cities. Also, the vast quantities of fresh water flowing from the melting glaciers into the oceans may cause a decrease in salt content; since certain climate-regulating ocean currents are a function of the water's salt content and temperature, they could be potentially slowed or shut off by these changes, which would lead to even more drastic climate change.

Global warming is affecting more than just the ocean's salinity, or salt content. As Bryn Nelson notes in "Human Prints on Warming," scientists have found that man's use of fossil fuels has led to an increase in ocean temperatures, which could put "pressure on water supplies in at-risk regions" and "endanger crucial warming cycles in the North Atlantic and force unpredictable and potentially catastrophic changes in the Arctic ecosystem."

A writer for the *UN Chronicle* in the next piece, "One Expedition's Story," describes the experiences of a team sent to the Himalayas by the United Nations Environment Programme (UNEP) to report on the ecological well-being of the region. What the expedition found alarmed them: The ice ranges and glaciers of the Himalayas have been melting at a fast pace, with the ice melt presenting a potential flood threat.

In number and intensity, the 2005 hurricane season was by far the most catastrophic in decades, if not centuries. During the emotional aftermath of Hurricanes Katrina and Rita, some were quick to attribute the violent weather to global warming. In "Global Warming: The Culprit?" Jeffrey Kluger explores the possible connection between climate change and the unprecedented number and magnitude of recent hurricanes. While few credible scientists claim that global warming is the cause of any particular hurricane, the storms are fueled by warm, tropical water; consequently, Kluger reports, some believe that rising ocean temperatures linked to climate change may account for the unexpected strength of recent storms.

Avi Salzman examines how the renowned autumn foliage of the northeastern United States may be losing some of its vibrant color due to increasing temperatures in the next entry, "A Season a Tad off Color, and Here's Why." The American Southwest is likewise feeling the impact of global warming, Juliet Eilperin reports in "Arid Arizona Points to Global Warming as Culprit." Prolonged droughts coupled with higher average temperatures have threatened the region's fragile ecosystem. Many believe that in the future, if current trends continue, the water needs of the Southwest's inhabitants will be difficult to accommodate.

The Pacific Northwest has not been immune to global warming either, as Craig Welch notes in "Climate Change Means Big Changes in Puget Sound." According to Welch, average winter temperatures have increased by 2.7 degrees Fahrenheit in the Northwest since 1950. Moreover, Puget Sound is rising and the mountain snowpack is melting earlier. These changes are already affecting the region's flora and fauna.

While Asian nations such as India and China have been rapidly industrializing over the last decade, their lax pollution controls and dependence on sometimes highly unrefined fossil fuels have caused the formation of virulent smog clouds. These clouds threaten to upset India's monsoon season and otherwise alter the climate, as Charles W. Petit notes in "A Darkening Sky," this chapter's final piece.

As the Arctic Melts, an Ancient Culture Faces Ruin

BY CHARLES WOHLFORTH
NATIONAL WILDLIFE, APRIL/MAY 2005

Richard Glenn stood as still as a statue, gazing through sun-glasses over the piercing white of the sea ice and the deep, flashing darkness of the Arctic Ocean. Three miles from shore north of Barrow, Alaska, he was a sentinel for a group of Iñupiaq subsistence hunters, watching for the spout of a bowhead whale but looking out even more carefully for signs that the ice edge upon which he was standing might break off and float away.

Glenn—geologist, ice scientist, founder of a research group called the Barrow Arctic Science Consortium and leader of an Iñupiaq whaling crew—knew from experience the dangers of an unexpected breakup. In spring 1997, he had been seven miles out on the frozen sea with his crew and more than 150 other whalers when the floes on which they stood began disintegrating. Polar bears moved in for a possible meal, and thick fog rose from the open water, making it impossible for helicopters from Barrow to find the crews. Whalers used the sound of the rotors to talk the choppers into position on handheld radios. The helicopters set down gingerly, half hovering so as not to break or flip the ice fragments. With people riding inside, and snow machines and sealskin boats dangling underneath, flights continued around the clock until the last hunter was safe ashore.

When I went whaling with Glenn's crew five years later, collecting material for a book on Arctic climate change, the ice had become even thinner and more unpredictable. Glenn and other captains ordered crews to pull back time and again, breaking camp to flee ahead of possible disaster. Then, one morning, he made the call to retreat a little sooner than the others, and his crew was the last to get off the ice before a great crack opened, sending adrift more than 100 whalers and a dead whale. Again the helicopters felt their way through fog and blowing snow. A heroic rescue brought all the people and most of their equipment home safely.

Something unprecedented had now happened twice in five years.

Article by Charles Wohlforth (*www.wohlforth.net*) from *National Wildlife* April/May 2005.
Copyright © Charles Wohlforth. Reprinted with permission.

Measures of Change

Climate scientists since the mid-1970s have predicted that warming would come first and strongest in the Arctic. For the past 15 years, the change has been increasingly evident, both to scientists and to indigenous people. In Alaska and western Canada, winter temperatures have risen as much as 7 degrees F over the past 50 years. Since the mid-1970s, the floating Arctic ice pack has lost an area the size of Texas and Arizona combined. With a shorter season of sea ice, fall storms batter Alaska's Arctic coast as never before, causing erosion that threatens communities. "We have talked about how change could have detrimental effects on our culture," Glenn says. "Longer distances and riskier journeys to get food. We're coastal people, and the coast is in danger."

Hunters remember how the ice used to be only a decade or two ago. In October, big, multiyear icebergs would run aground off the gravel shore. Flat new ice would form one cold night to connect the icebergs, which would anchor and strengthen the entire ice sheet the way a steel frame holds the skin of a skyscraper. By the time the sun returned from a two-month absence in late January, the ice would have become an extension of the land, a reliable place for pursuing a family's sustenance until the ice broke up sometime in June or July.

As the Arctic environment changes, the centuries-old Iñupiaq culture will be forced to change as well.

Now the multiyear icebergs do not appear in fall, so the shore ice is far less stable. It freezes later and breaks up earlier. The Iñupiat cannot count on it for the spring hunt as they once could. "If we start losing the spring season, we have to totally rethink ice safety," Glenn says. "Things that were true for fathers won't be true for sons."

As the Arctic environment changes, the centuries-old Iñupiaq culture will be forced to change as well, perhaps becoming in some ways as unrecognizable as the new landscape. The elders, who hold the greatest store of old knowledge, were the first to report the unprecedented ecological changes. Thomas Itta, Sr., in his seventies, perceives these changes whenever he goes hunting on the flat tundra around his village of Atqasuk. He sees the accelerated growth of bushy willows and how the shrubs have redistributed the snowpack. He hears the cries of invading sea gulls that have taken the place of some disappearing waterfowl. He feels the mosquitoes on his skin a month earlier in summer. He tastes a change in caribou meat due to an explosion of tundra flowers, on which the animals graze. The very geography of Itta's world is changing: As the permafrost softens near the surface, ponds are connecting with other water bodies through formerly solid shores. "These lakes and ponds, they never used to break into the others or break into the river," he says. "Starting in the nineties, they started breaking."

Seeking Solutions

Does it matter where Arctic lakes find their boundaries or how snow piles up in the willows that have sprouted from the tundra?

Scientists say it matters a lot. New research indicates that a warming Arctic can accelerate climate change worldwide. Arctic snow reflects the warmth of the sun back to space. As winter becomes shorter, and snow is redistributed by brush, the ground absorbs ever more heat. Shrubs also absorb more heat in the summertime than does the flat tundra. The thawing soil underneath emits carbon stored for millennia in dead organic matter. Itta's changing lake shores are a signal of that soil thawing.

Climate change affects Arctic people's lives every day. The melting of permafrost, the permanently frozen layer under much of Alaska, has turned solid ground into mush in some places, collapsing roads and even causing supports to tip under some parts of the trans-Alaska oil pipeline. In Barrow, changes in ocean waves have washed away ancient archaeological sites, and the bluff that overlooks the sea has crept perilously close to houses. A decision must be made either to move the houses or to defend the coast from waves. Other villages, such as Shishmaref, which sits on a shifting barrier island, have fewer options and less time to decide their future: either disperse their ancient tribe, or move to a new location at enormous cost.

New research indicates that a warming Arctic can accelerate climate change worldwide.

Moving houses is one way of adapting to a changing climate. Another is to change hunting practices. As whaling has become more dangerous on spring sea ice, Barrow whalers have hunted more in autumn, when ice is absent and motorboats are more practical than traditional skin boats. As the warming climate causes autumn waves to rise, Iñupiaq whalers are buying bigger boats with bigger motors to stay safe in pursuit of the 56 whales that the International Whaling Commission has allocated to them yearly.

When working on my book, I sat for weeks with whaling crews on the spring ice while warm weather spoiled their frozen meat, created puddles on the ice trail and brought dispiriting rain that melted the snow. But even in that terrible spring we landed a 54-foot bowhead whale. Some 200 villagers gathered to pull the 100-ton animal from the water with ropes and muscle.

The following fall, I joined Richard Glenn's crew once again. The air of late September on the Arctic Ocean felt even colder than in winter—the dampness cut through layers of down. We patrolled, scanning the horizon for a whale's spout, but the waves were high. Later I learned that the sea ice had shrunk more that season than ever before, giving waves lots of room to build. Most boats went home, because even if a crew managed to take a whale, towing it ashore would be impossible in those seas.

As we rounded Point Barrow, the northernmost point in the United States, I saw a group of polar bears. Several dozen were stranded in the area, having landed from sea ice that then withdrew hundreds of miles from shore. Through the fog and wind, I could see big white breakers tumbling ashore on the low gravel. The bears saw us as well, and one reared up to watch the boat go by. Another charged to the water as if to attack the tumbling waves that had trapped it there.

A Bit of Icy Antarctica Is Sliding Toward the Sea

By Richard A. Kerr
Science, September 29, 2004

The latest gauging of West Antarctic glaciers confirms that when the ocean eats at one end of a glacier, it can draw far-distant ice toward the sea, with potentially dangerous consequences

As the global climate warms up, glaciologists' big worry is polar ice, especially the ice sheet of West Antarctica, the muscular arm that juts from the huge mound of ice in East Antarctica. They aren't concerned about warmer air per se; even the thinner West Antarctic Ice Sheet (WAIS) would hold out against its effects for millennia. But researchers have long wondered whether warming could somehow get at the WAIS indirectly, destabilize it, and send its ice into the sea to melt, raising sea level up to a disastrous 5 meters in a few centuries. With the publication online (www.sciencemag.org/cgi/content/abstract/1099650) of the latest survey of glaciers flowing into West Antarctica's Amundsen Sea, most glaciologists now allow that there probably is a way for warming to accelerate the movement of at least some of the WAIS ice toward the sea.

Glaciologist Robert Thomas of NASA contractor EG&G at the Wallops Island facility in Virginia and colleagues confirm that the half-dozen glaciers flowing into the Amundsen Sea have been getting thinner and thinner the past 15 years, and that one of them—the Pine Island Glacier—has been flowing faster and faster for more than 100 kilometers inland. "It's not necessarily a sign of [WAIS] collapse," says glaciologist Richard Alley of Pennsylvania State University, University Park, "but it could lead to a collapse."

However, no one can say whether the recent glacial acceleration will continue, whether it could reach more distant ice if it does continue, or whether other, more voluminous parts of the WAIS could suffer similar effects. "We're not running for the hills," says Alley, but "this is the wake-up call for the scientific community to get serious about it all."

Since the start of the 1990s, glaciologists have been closely monitoring the flow of ice from the Pine Island Glacier and nearby glaciers using motion-sensing radar, ice-penetrating radar, and laser and radar altimeters mounted on satellites and aircraft. By the end of the decade, the ice in at least some glacial channels nearing the sea seemed to be thinning and accelerating.

To learn more, Thomas and his colleagues, in cooperation with Centra de Hstudios Cientificos in Valdivia, Chile, rode an instrument-laden Chilean Navy P-3 aircraft 2700 kilometers to the remote Amundsen Sea coast. The onboard ice-penetrating radar found that the ice is far thicker than thought, on average 400 meters deeper than previously estimated near the coast. Combined with satellite radar velocity estimates from the late 1990s, those greater thicknesses implied that the glaciers are hauling away about 253 cubic kilometers of ice per year. That's about 90 cubic kilometers more than accumulates each year from snowfall.

By analyzing recent satellite radar data, Thomas and colleagues confirm that ice withdrawals have been accelerating, at least through the Pine Island Glacier, the largest of the group. They calculate that it sped up by 3.5% between April 2001 and early 2003, making for a 25% increase since the mid-1970s. And the drawdown

---

For 30 years, glaciologists have debated whether one part of a glacier can "feel" what's happening in a distant part of the same glacier.

---

is not limited to areas near the coast. The P-3 data show a thinning, presumably induced by the faster flow, that extends along the main trunk of the Pine Island Glacier and averages about 1.2 meters per year between 100 and 300 kilometers inland.

These latest results from West Antarctica confirm an unsettling view of glacier behavior. For 30 years, glaciologists have debated whether one part of a glacier can "feel" what's happening in a distant part of the same glacier. At the coastal end of the Pine Island Glacier, for example, warmer water seems to be melting the underside of the glacier's floating ice shelf (*Science*, 24 July 1998, pp. 499 and 549), pushing landward the point at which the advancing glacier floats off the sea floor.

If an ice shelf pinned against an embayment's shore and floor helps slow a glacier's flow—as was hypothesized in the 1970s—and if changes at the coast could make themselves felt far up the glacier, then the Pine Island Glacier's so-called grounding line retreat would accelerate glacier flow well upstream. The researchers think that's what they're seeing. "I'm convinced the glacier feels what is happening a long way away," says Thomas. Similar accelerations struck after two other floating ice tongues recently broke up in West Antarctica and Greenland (*Science*, 30 August 2002, p. 1494).

"It's a very impressive piece of work," says Alley. "Too many different lines of evidence are agreeing now" for them to be wrong about the thinning or the speedup of the past 10 to 15 years. "Ice shelves may well play a role in the dynamics of glaciers," agrees geoscientist Michael Oppenheimer of Princeton University in New Jersey. But the next problem is that "we don't know why things are melting away at Pine Island Glacier." Oceanographers can't say whether the ocean warming that seems responsible is part of a cycle that will reverse itself or a long-term trend driven by greenhouse warming. And they can't say whether the WAIS's two largest ice shelves—the Texas-size Ronne and Ross ice shelves—could be melted as well.

Even if glaciologists knew what the ocean was going to do, their models for predicting glacier behavior are still so rudimentary that they can't say whether more distant, slower moving ice feeding the main ice streams will respond too. So plenty of uncertainties remain, notes Oppenheimer, but he adds, "I'm starting to get worried."

The Big Thaw

By Daniel Glick
National Geographic, September 2004

"If we don't have it, we don't need it," pronounces Daniel Fagre as we throw on our backpacks. We're armed with crampons, ice axes, rope, GPS receivers, and bear spray to ward off grizzlies, and we're trudging toward Sperry Glacier in Glacier National Park, Montana. I fall in step with Fagre and two other research scientists from the U.S. Geological Survey Global Change Research Program.

They're doing what they've been doing for more than a decade: measuring how the park's storied glaciers are melting.

So far, the results have been positively chilling. When President Taft created Glacier National Park in 1910, it was home to an estimated 150 glaciers. Since then the number has decreased to fewer than 30, and most of those remaining have shrunk in area by two-thirds. Fagre predicts that within 30 years most if not all of the park's namesake glaciers will disappear.

"Things that normally happen in geologic time are happening during the span of a human lifetime," says Fagre. "It's like watching the Statue of Liberty melt."

Scientists who assess the planet's health see indisputable evidence that Earth has been getting warmer, in some cases rapidly. Most believe that human activity, in particular the burning of fossil fuels and the resulting buildup of greenhouse gases in the atmosphere, have influenced this warming trend. In the past decade scientists have documented record-high average annual surface temperatures and have been observing other signs of change all over the planet: in the distribution of ice, and in the salinity, levels, and temperatures of the oceans.

"This glacier used to be closer," Fagre declares as we crest a steep section, his glasses fogged from exertion. He's only half joking. A trailside sign notes that since 1901, Sperry Glacier has shrunk from more than 800 acres to 300 acres. "That's out of date," Fagre says, stopping to catch his breath. "It's now less than 250 acres."

Everywhere on Earth ice is changing. The famed snows of Kilimanjaro have melted more than 80 percent since 1912. Glaciers in the Garhwal Himalaya in India are retreating so fast that researchers believe that most central and eastern Himalayan glaciers could virtually disappear by 2035. Arctic sea ice has thinned significantly over the past half century, and its extent has declined by about 10 percent in the past 30 years. NASA's repeated laser altimeter readings show the edges of Greenland's ice sheet shrinking. Spring freshwater ice breakup in the Northern Hemisphere now occurs nine days earlier than it did 150 years ago, and autumn freezeup 10 days later. Thawing permafrost has caused the ground to subside more than 15 feet in parts of Alaska. From the Arctic to Peru, from Switzerland to the equatorial glaciers of Irian Jaya in Indonesia, massive ice fields, monstrous glaciers, and sea ice are disappearing, fast.

Driving around Louisiana's Gulf Coast, Windell Curole can see the future, and it looks pretty wet.

When temperatures rise and ice melts, more water flows to the seas from glaciers and ice caps, and ocean water warms and expands in volume. This combination of effects has played the major role in raising average global sea level between four and eight inches in the past hundred years, according to the Intergovernmental Panel on Climate Change (IPCC).

Scientists point out that sea levels have risen and fallen substantially over Earth's 4.6-billion-year history. But the recent rate of global sea level rise has departed from the average rate of the past two to three thousand years and is rising more rapidly—about one-tenth of an inch a year. A continuation or acceleration of that trend has the potential to cause striking changes in the world's coastlines.

Driving around Louisiana's Gulf Coast, Windell Curole can see the future, and it looks pretty wet. In southern Louisiana coasts are literally sinking by about three feet a century, a process called subsidence. A sinking coastline and a rising ocean combine to yield powerful effects. It's like taking the global sea-level-rise problem and moving it along at fast-forward.

The seventh-generation Cajun and manager of the South Lafourche Levee District navigates his truck down an unpaved mound of dirt that separates civilization from inundation, dry land from a swampy horizon. With his French-tinged lilt, Curole points to places where these bayous, swamps, and fishing villages portend a warmer world: his high school girlfriend's house partly submerged, a cemetery with water lapping against the white tombs, his grandfather's former hunting camp now afloat in a stand of skeleton oak snags. "We live in a place of almost land, almost water," says the 52-year-old Curole.

Rising sea level, sinking land, eroding coasts, and temperamental storms are a fact of life for Curole. Even relatively small storm surges in the past two decades have overwhelmed the system of dikes, levees, and pump stations that he manages, upgraded in the

1990s to forestall the Gulf of Mexico's relentless creep. "I've probably ordered more evacuations than any other person in the country," Curole says.

The current trend is consequential not only in coastal Louisiana but around the world. Never before have so many humans lived so close to the coasts: More than a hundred million people worldwide live within three feet of mean sea level. Vulnerable to sea-level rise, Tuvalu, a small country in the South Pacific, has already begun formulating evacuation plans. Megacities where human populations have concentrated near coastal plains or river deltas—Shanghai, Bangkok, Jakarta, Tokyo, and New York—are at risk. The projected economic and humanitarian impacts on low-lying, densely populated, and desperately poor countries like Bangladesh are potentially catastrophic. The scenarios are disturbing even in wealthy countries like the Netherlands, with nearly half its landmass already at or below sea level.

Rising sea level produces a cascade of effects. Bruce Douglas, a coastal researcher at Florida International University, calculates that every inch of sea-level rise could result in eight feet of horizontal retreat of sandy beach shorelines due to erosion. Furthermore,

---

In some places marvels of human engineering worsen effects from rising seas in a warming world.

---

when salt water intrudes into freshwater aquifers, it threatens sources of drinking water and makes raising crops problematic. In the Nile Delta, where many of Egypt's crops are cultivated, widespread erosion and saltwater intrusion would be disastrous—since the country contains little other arable land.

In some places marvels of human engineering worsen effects from rising seas in a warming world. The system of channels and levees along the Mississippi effectively stopped the millennia-old natural process of rebuilding the river delta with rich sediment deposits. In the 1930s, oil and gas companies began to dredge shipping and exploratory canals, tearing up the marshland buffers that helped dissipate tidal surges. Energy drilling removed vast quantities of subsurface liquid, which studies suggest increased the rate at which the land is sinking. Now Louisiana is losing approximately 25 square miles of wetlands every year, and the state is lobbying for federal money to help replace the upstream sediments that are the delta's lifeblood.

Local projects like that might not do much good in the very long run, though, depending on the course of change elsewhere on the planet. Part of Antarctica's Larsen Ice Shelf broke apart in early 2002. Although floating ice does not change sea level when it melts (any more than a glass of water will overflow when the ice cubes in it melt), scientists became concerned that the collapse could fore-shadow the breakup of other ice shelves in Antarctica and allow increased glacial discharge into the sea from ice sheets on the continent. If the West Antarctic ice sheet were to break up, which scientists consider very unlikely this century, it alone contains enough ice to raise sea level by nearly 20 feet.

Even without such a major event, the IPCC projected in its 2001 report that sea level will rise anywhere between 4 and 35 inches by the end of the century. The high end of that projection—nearly three feet—would be "an unmitigated disaster," according to Douglas.

Down on the bayou, all of those predictions make Windell Curole shudder. "We're the guinea pigs," he says, surveying his aqueous world from the relatively lofty vantage point of a 12-foot-high earthen berm. "I don't think anybody down here looks at the sea-level-rise problem and puts their heads in the sand." That's because soon there may not be much sand left.

Rising sea level is not the only change Earth's oceans are undergoing. The ten-year-long World Ocean Circulation Experiment, launched in 1990, has helped researchers to better understand what is now called the ocean conveyor belt.

Oceans, in effect, mimic some functions of the human circulatory system.

Oceans, in effect, mimic some functions of the human circulatory system. Just as arteries carry oxygenated blood from the heart to the extremities, and veins return blood to be replenished with oxygen, oceans provide life-sustaining circulation to the planet. Propelled mainly by prevailing winds and differences in water density, which changes with the temperature and salinity of the seawater, ocean currents are critical in cooling, warming, and watering the planet's terrestrial surfaces—and in transferring heat from the Equator to the Poles.

The engine running the conveyor belt is the density-driven thermohaline circulation "thermo" for heat and "haline" for salt). Warm, salty water flows from the tropical Atlantic north toward the Pole in surface currents like the Gulf Stream. This saline water loses heat to the air as it is carried to the far reaches of the North Atlantic. The coldness and high salinity together make the water more dense, and it sinks deep into the ocean. Surface water moves in to replace it. The deep, cold water flows into the South Atlantic, Indian, and Pacific Oceans, eventually mixing again with warm water and rising back to the surface.

Changes in water temperature and salinity, depending on how drastic they are, might have considerable effects on the ocean conveyor belt. Ocean temperatures are rising in all ocean basins and at much deeper depths than previously thought, say scientists at the National Oceanic and Atmospheric Administration (NOAA). Arguably, the largest oceanic change ever measured in the era of modern instruments is in the declining salinity of the subpolar seas bordering the North Atlantic.

Robert Gagosian, president and director of the Woods Hole Oceanographic Institution, believes that oceans hold the key to potential dramatic shifts in the Earth's climate. He warns that too much change in ocean temperature and salinity could disrupt the North Atlantic thermohaline circulation enough to slow down or possibly halt the conveyor belt—causing drastic climate changes in time spans as short as a decade.

The future breakdown of the thermohaline circulation remains a disturbing, if remote, possibility. But the link between changing atmospheric chemistry and the changing oceans is indisputable, says Nicholas Bates, a principal investigator for the Bermuda Atlantic Time-series Study station, which monitors the temperature, chemical composition, and salinity of deep-ocean water in the Sargasso Sea southeast of the Bermuda Triangle.

Oceans are important sinks, or absorption centers, for carbon dioxide, and take up about a third of human-generated CO_2. Data from the Bermuda monitoring programs show that CO_2 levels at the ocean surface are rising at about the same rate as atmospheric CO_2. But it is in the deeper levels where Bates has observed even greater change. In the waters between 250 and 450 meters (820 and 1,476 feet) deep, CO_2 levels are rising at nearly twice the rate as in the surface waters. "It's not a belief system; it's an observable scientific fact," Bates says. "And it shouldn't be doing that unless something fundamental has changed in this part of the ocean."

While scientists like Bates monitor changes in the oceans, others evaluate CO_2 levels in the atmosphere. In Vestmannaeyjar, Iceland, a lighthouse attendant opens a large silver suitcase that looks like something out of a James Bond movie, telescopes out an attached 15-foot rod, and flips a switch, activating a computer that controls several motors, valves, and stopcocks. Two two-and-a-half-liter flasks in the suitcase fill with ambient air. In North Africa, an Algerian monk at Assekrem does the same. Around the world, collectors like these are monitoring the cocoon of gases that compose our atmosphere and permit life as we know it to persist.

When the weekly collection is done, all the flasks are sent to Boulder, Colorado. There, Pieter Tans, a Dutch-born atmospheric scientist with NOAA's Climate Monitoring and Diagnostics Laboratory, oversees a slew of sensitive instruments that test the air in the flasks for its chemical composition. In this way Tans helps assess the state of the world's atmosphere.

By all accounts it has changed significantly in the past 150 years.

Walking through the various labs filled with cylinders of standardized gas mixtures, absolute manometers, and gas chromatographs, Tans offers up a short history of atmospheric monitoring. In the late 1950s a researcher named Charles Keeling began measuring CO_2 in the atmosphere above Hawaii's 13,679-foot Mauna Loa. The first thing that caught Keeling's eye was how CO_2 level rose and fell seasonally. That made sense since, during spring and summer, plants take in CO_2 during photosynthesis and produce oxygen in the atmosphere. In the fall and winter, when plants decay, they release greater quantities of CO_2 through respiration and decay. Keeling's vacillating seasonal curve became famous as a visual representation of the Earth "breathing."

Something else about the way the Earth was breathing attracted Keeling's attention. He watched as CO_2 level not only fluctuated seasonally, but also rose year after year. Carbon dioxide level has climbed from about 315 parts per million (ppm) from Keeling's first readings in 1958 to more than 375 ppm today. A primary source for this rise is indisputable: humans' prodigious burning of carbon-laden fossil fuels for their factories, homes, and cars.

Tans shows me a graph depicting levels of three key greenhouse gases—CO_2, methane, and nitrous oxide—from the year 1000 to the present. The three gases together help keep Earth, which would otherwise be an inhospitably cold orbiting rock, temperate by orchestrating an intricate dance between the radiation of heat from Earth back to space (cooling the planet) and the absorption of radiation in the atmosphere (trapping it near the surface and thus warming the planet).

Tans and most other scientists believe that greenhouse gases are at the root of our changing climate. "These gases are a climate-change driver," says Tans, poking his graph definitively with his index finger. The three lines on the graph follow almost identical patterns: basically flat until the mid-1800s, then all three move upward in a trend that turns even more sharply upward after 1950. "This is what we did," says Tans, pointing to the parallel spikes. "We have very significantly changed the atmospheric concentration of these gases. We know their radiative properties," he says. "It is inconceivable to me that the increase would not have a significant effect on climate."

Exactly how large that effect might be on the planet's health and respiratory system will continue to be a subject of great scientific and political debate—especially if the lines on the graph continue their upward trajectory.

Eugene Brower, an Inupiat Eskimo and president of the Barrow Whaling Captains' Association, doesn't need fancy parts-per-million measurements of CO_2 concentrations or long-term sea-level gauges to tell him that his world is changing.

"It's happening as we speak," the 56-year-old Brower says as we drive around his home in Barrow, Alaska—the United States' northernmost city—on a late August day. In his fire chief's truck, Brower takes me to his family's traditional ice cellars, painstakingly dug into the permafrost, and points out how his stores of muktuk—whale skin and blubber—recently began spoiling in the fall because melting water drips down to his food stores. Our next stop is the old Bureau of Indian Affairs school building. The once impenetrable permafrost that kept the foundation solid has bucked and heaved so much that walking through the school is almost like walking down the halls of an amusement park fun house. We head to the eroding beach and gaze out over open water. "Normally by now the ice would be coming in," Brower says, scrunching up his eyes and scanning the blue horizon.

We continue our tour. Barrow looks like a coastal community under siege. The ramshackle conglomeration of weather-beaten houses along the seaside gravel road stands protected from fall storm surges by miles-long berms of gravel and mud that block views of migrating gray whales. Yellow bulldozers and graders patrol the coast like sentries.

The Inupiat language has words that describe many kinds of ice. *Piqaluyak* is salt-free multiyear sea ice. *Ivuniq* is a pressure ridge. *Sarri* is the word for pack ice, *tuvaqtaq* is bottom-fast ice, and shore-fast ice is *tuvaq*. For Brower, these words are the currency of hunters who must know and follow ice patterns to track bearded seals, walruses, and bowhead whales.

There are no words, though, to describe how much, and how fast, the ice is changing. Researchers long ago predicted that the most visible impacts from a globally warmer world would occur first at high latitudes: rising air and sea temperatures, earlier snowmelt, later ice freeze-up, reductions in sea ice, thawing permafrost, more erosion, increases in storm intensity. Now all those impacts have been documented in Alaska. "The changes observed here provide an early warning system for the rest of the planet," says Amanda Lynch, an Australian researcher who is the principal investigator on a project that works with Barrow's residents to help them incorporate scientific data into management decisions for the city's threatened infrastructure.

Before leaving the Arctic, I drive to Point Barrow alone. There, at the tip of Alaska, roughshod hunting shacks dot the spit of land that marks the dividing line between the Chukchi and Beaufort Seas. Next to one shack someone has planted three eight-foot sticks of white driftwood in the sand, then crisscrossed their tops with whale baleen, a horny substance that whales of the same name use to filter life-sustaining plankton out of seawater. The baleen, curiously, looks like palm fronds.

So there, on the North Slope of Alaska, stand three makeshift palm trees. Perhaps they are no more than an elaborate Inupiat joke, but these Arctic palms seem an enigmatic metaphor for the Earth's future.

Human Prints on Warming

By Bryn Nelson
Newsday, February 20, 2005

Scientists have announced what they say is stunning evidence of ocean warming linked to human activity over the past 40 years.

The computer model–aided finding, if corroborated, may add a major body of evidence to support the global warming concept backed by most scientists, since oceans are known to be major drivers in the world's atmospheric conditions.

"The debate is no longer, 'Is there a global warming signal?' The question is, what are we going to do about it?" said Tim Barnett, a research marine physicist at Scripps Institution of Oceanography in La Jolla, Calif., and study co-author. Barnett, who presented the research Friday at the annual conference of the American Association for the Advancement of Science, said he and colleagues were stunned by the results, which drew gasps from the audience.

Besides placing additional pressure on water supplies in at-risk regions such as the Southwest, the South American Andes, and western China in the next few decades, the ocean warming trend may endanger crucial warming cycles in the North Atlantic and force unpredictable and potentially catastrophic changes in the Arctic ecosystem, other scientists reported at the same session.

Barnett and other scientists ended talks with pleas for reduced emissions of carbon dioxide, a major component of greenhouse gases that most link to global warming.

Temperature Readings

The research arrives the same week as the implementation of the Kyoto accord's initial program aimed at curbing these emissions. The treaty has been ratified by 141 nations but opposed by a handful of countries including the United States, which has questioned the treaty's underlying science and objected to provisions deemed harmful to its economic interests.

Computer models of climate have been routinely attacked by skeptics as unreliable, in part because scientists say the ocean's crucial role hasn't been adequately considered.

"The real action is in the ocean," Barnett said during a Thursday news preview, defending his study as more robust since it relies on a model with "unprecedented resolution" over the ocean and on wide-ranging water temperature measurements. He said the temperature readings, collected from varying depths in six Atlantic,

Pacific, and Indian Ocean regions between 1960 and 1999, yielded a unique fingerprint that could be matched against a list of "the usual suspects" blamed for the recent warming phenomenon.

Explanations such as natural variability or changes in solar radiation and volcanic activity, Barnett said, produced poor fits, showing graphs to back his claim. But the model of human-related warming produced a signature "almost identical" to the real temperature measurements in each ocean site over the same period, he said.

"It's very compelling. It's nothing we ever expected."

Independent Studies

Independent models produced by scientists at the Hadley Centre for Climate Prediction and Research in the United Kingdom yielded similar, though somewhat more variable results, based on a set of graphs presented by Barnett.

Ruth Curry, a physical oceanographer at the Woods Hole Oceanographic Institution, said a global warming–linked loss of ice around the world, especially in the Arctic, is upsetting the Earth's ocean water cycles that affect everything from evaporation to precipitation.

Since the 1960s, according to one of her studies now in review, about 4,800 cubic miles of extra fresh water have dumped into the upper reaches of the North Atlantic. Curry said the region includes two sensitive spots along the ocean's conveyor belt–like system that circulates warmer water from lower latitudes to upper latitudes, warming the air as well. Too much freshwater accumulation at key locations could disrupt the entire circulation system, leading to slowdown or shutdown of the conveyor belt and its accompanying heat, she said. The range of atmospheric consequences could include drought in some parts of the world and a significant cooling of northern Europe, though a new Ice Age is unlikely.

Humans aren't the only ones in harm's way, other researchers said. University of Miami marine biologist Sharon Smith said loss of the Arctic's ice cover could be catastrophic for animals such as walruses and polar bears.

Smith also found a mass die-off of gull-like birds known as short-tailed shearwaters in 1997 was linked to the proliferation of an unusual plant in the Bering Sea. Warming conditions led to an enormous bloom of the one-celled marine plant, which discolored the water and prevented birds from seeing prey below the surface. Hundreds of thousands of birds starved to death as a result, she said.

One Expedition's Story

UN CHRONICLE, SEPTEMBER/NOVEMBER 2002

An expedition, dispatched to the Himalayas to chronicle the environmental health of one of the world's most famous mountain ranges, has gathered startling evidence of the impacts of climate change. Backed by the UN Environment Programme (UNEP), the team has learned that the glacier from where Sir Edmund Hillary and Tenzing Norgay set out to conquer Everest nearly fifty years ago has retreated up the mountain by around five kilometres. Roger Payne, Sports and Development Director at the International Mountaineering and Climbing Federation (UIAA) and one of the expedition's leaders, said: "It is clear that global warming is emerging as one of if not the biggest threat to mountain areas. The evidence of climate change was all around us, from huge scars gouged in the landscapes by sudden glacial floods to the lakes swollen by melting glaciers. But it is the observations of some of the people we met, many of whom have lived in the area all their lives, that really hit home."

The seven-strong expedition, which set out from Kathmandu on 16 May 2002, returned on 1 June after climbing Island Peak, which is 6,189 metres (20,305 feet) above sea level in the Khumbu region of Nepal. It visited the famous Thyangboche Monastery and talked to experts, including those in the Sagarmatha (Everest) National Park. It was in conversation with Tashi Janghu Sherpa, President of the Nepal Mountain Association, that the team first learned of rising concern among local people over the impacts of global warming.

Ian McNaught-Davis, President of UIAA and another expedition leader, said: "He told us that he had seen quite rapid and significant changes over the past twenty years in the ice fields and that these changes appeared to be accelerating. He told us that Hillary and Tenzing would now have to walk two hours to find the edge of the glacier, which was close to their original base camp in 1953, which means that it has retreated by between four and six kilometres. And that around Island Peak, so called because it once resembled an island in a sea of ice, there was once a network of small ponds. Today, they have merged into a big, several kilometre-long lake as a result of the glaciers melting. Mr. Janghu said he was worried—

worried that glaciers would continue shrinking and that the melt waters would trigger floods, sending huge quantities of water, rubble and mud down the valley."

At the Thyangboche Monastery, home to sixty Buddhist monks, the team met with Lama Rinpoche who has lived there for over thirty years and witnessed two big floods, the result of melting glaciers causing local lakes to burst. One recent flood had washed away the old wooden bridges downstream. New metal ones have been built higher and 100 metres longer, replacing the older 50-metre ones, to try and reduce the chances of similar damage from a future flood. "It was the Lama's impression that such events were becoming more frequent and a rising phenomenon of the past eight to nine years," said Mr. McNaught-Davis.

There has been concern that rising numbers of tourists and climate change might also be having impacts on the vegetation of the area. An estimated 27,000 people a year visit the area, up from a handful in the early 1960s. Tourists now outnumber the local sherpa population, which totals 3,000 in the Khumbu region of Nepal's Solu Khumbu District.

Julia-Ann Clyma, another member of the expedition from New Zealand, said that just below the village of Thyangboche people were developing a medicinal herb garden in an attempt to preserve local medicinal plants and knowledge. "We saw a lot of impressive efforts by local people to make themselves less dependent on food imports, including the development of greenhouse crops and fruit orchards," she said.

The team was also impressed by the numerous reafforestation schemes under way, aimed at balancing the fuel wood needs of local people and tourists with the need to maintain healthy forests. This appears in line with the research of the Mountain Institute, which found that forest cover below the snow and ice line "remains essentially unchanged from the 1950s. Natural forest regeneration appears to be increasing in many areas, and tree growth in the vicinity of the Namche Bazaar and other villages has increased as a result of successful plantation efforts over the past 15 years." But above 4,000 metres, over-harvesting of high altitude juniper shrubs and cushion plants for fuel, nearly all of which is tourist-related, is having a serious impact on the environment. These impacts include erosion and loss of wildlife.

However, local community action groups are being developed to restore these degraded habitats. Plans include banning the harvesting of alpine shrubs and the development of subsidies to encourage the sustainable exploitation of trees, such as the plentiful supplies of birch and rhododendrons from lower down. Building shelters for porters at major trekking villages is also under discussion. Currently, many porters sleep outside and burn wood to keep warm.

Pemba Geljen Sherpa, the expedition's guide who has lived in the area all his life, said he had witnessed dramatic changes in his lifetime. Traditional dress and customs were fast disappearing, but he suggested this was an inevitable consequence of the modern world. "It is all changing, you do not see the same traditional dancing or singing of my parents' generation." But he rejected suggestions that tourism should be curtailed: "We need more, not less, tourism here to boost the economy and give people jobs, incomes and education. I think we can manage it, so that it is the right kind of tourism that respects local people and local landscapes. What we cannot control is global warming—that is in the hands of others. We here in Nepal produce tiny amounts of the gases linked with global warming. It is up to the big, industrial countries of Europe, North America and Japan to act to save our mountains and the environment upon which our livelihoods depend."

Global Warming: The Culprit?

By Jeffrey Kluger
TIME, October 3, 2005

Nature doesn't always know when to quit—and nothing says that quite like a hurricane. The atmospheric convulsion that was Hurricane Katrina had barely left the Gulf Coast before its sister Rita was spinning to life out in the Atlantic. In the three weeks between them, five other named storms had lived and died in the warm Atlantic waters without making the same headlines their ferocious sisters did. With more than two months left in the official hurricane season, only Stan, Tammy, Vince and Wilma are still available on the National Hurricane Center's annual list of 21 storm names. If the next few weeks go like the past few, those names will be used up too, and the storms that follow will be identified simply by Greek letters. Never in the 52 years we have been naming storms has there been a Hurricane Alpha.

If 2005 goes down as the worst hurricane season on record in the North Atlantic, it will join 2004 as one of the most violent ever. And these two seasons are part of a trend of increasingly powerful and deadly hurricanes that has been playing out for more than 10 years. Says climatologist Judy Curry, chair of the School of Earth and Atmospheric Sciences at the Georgia Institute of Technology: "The so-called once-in-a-lifetime storm isn't even once in a season anymore."

Head-snapping changes in the weather like this inevitably raise the question, Is global warming to blame? For years, environmentalists have warned that one of the first and most reliable signs of a climatological crash would be an upsurge in the most violent hurricanes, the kind that thrive in a suddenly warmer world. Scientists are quick to point out that changes in the weather and climate change are two different things. But now, after watching two Gulf Coast hurricanes reach Category 5 in the space of four weeks, even skeptical scientists are starting to wonder whether something serious might be going on.

"There is no doubt that climate is changing and humans are partly responsible," says Kevin Trenberth, head of the climate-analysis section at the National Center for Atmospheric Research (NCAR) in Boulder, Colo. "The odds have changed in favor of more intense storms and heavier rainfalls." Says NCAR meteorologist Greg Holland: "These are not small changes. We're talking about a very large change."

But do scientists really know for sure? Can man-made greenhouse gases really be blamed for the intensity of storms like Rita and Katrina? Or are there, as other experts insist, too many additional variables to say one way or the other?

That global warming ought to, in theory, exacerbate the problem of hurricanes is an easy conclusion to reach. Few scientists doubt that carbon dioxide and other greenhouse gases raise the temperature of Earth's atmosphere. Warmer air can easily translate into warmer oceans—and warm oceans are the jet fuel that drives the hurricane's turbine. When Katrina hit at the end of August, the Gulf of Mexico was a veritable hurricane refueling station, with water up to 5°F higher than normal. Rita too drew its killer strength from the Gulf, making its way past southern Florida as a Category 1 storm, then exploding into a Category 5 as it moved westward. "The Gulf is really warm this year, and it's just cooking those tropical storms," says Curry.

> *That global warming ought to, in theory, exacerbate the problem of hurricanes is an easy conclusion to reach.*

Local hot spots like this are not the same as global climate change, but they do appear to be part of a larger trend. Since 1970, mean ocean surface temperatures worldwide have risen about 1°F. Those numbers have moved in lockstep with global air temperatures, which have also inched up a degree. The warmest year ever recorded was 1998, with 2002, 2003 and 2004 close behind it.

So that ought to mean a lot more hurricanes, right? Actually, no—which is one of the reasons it's so hard to pin these trends down. The past 10 stormy years in the North Atlantic were preceded by many very quiet ones—all occurring at the same time that global temperatures were marching upward. Worldwide, there's a sort of equilibrium. When the number of storms in the North Atlantic increases, there is usually a corresponding fall in the number of storms in, say, the North Pacific. Over the course of a year, the variations tend to cancel one another out. "Globally," says atmospheric scientist Kerry Emanuel of the Massachusetts Institute of Technology, "we do not see any increase at all in the frequency of hurricanes."

But frequency is not the same as intensity, and two recent studies demonstrate that difference. Two weeks ago, a team of scientists that included Curry and Holland published a study in the journal *Science* that surveyed global hurricane frequency and intensity over the past 35 years. On the whole, they found, the number of Category 1, 2 and 3 storms has fallen slightly, while the number of Categories 4 and 5 storms—the most powerful ones—has climbed dramatically. In the 1970s, there were an average of 10 Category 4 and 5 hurricanes a year worldwide. Since 1990, the annual number has nearly doubled, to 18. Overall, the big storms have grown from just 20% of the global total to 35%. "We have a sustained increase [in hurricane intensity] over 30 years all over the globe," says Holland.

Emanuel came at the same question differently but got the same results. In a study published in the journal *Nature* last month, he surveyed roughly 4,800 hurricanes in the North Atlantic and North Pacific over the past 56 years. While he too found no increase in the total number of hurricanes, he found that their power—measured by wind speed and duration—had jumped 50% since the mid-1970s. "The storms are getting stronger," Emanuel says, "and they're lasting longer."

Several factors help feed the trend. For example, when ocean temperatures rise, so does the amount of water vapor in the air. A moister atmosphere helps fuel storms by giving them more to spit out in the form of rain and by helping drive the convection that gives them their lethal spin. Warm oceans produce higher levels of vapor than cool oceans—at a rate of about 1.3% more per decade since 1988, according to one study—and nothing gets that process going better than greenhouse-heated air. "Water vapor increases the rainfall intensity," says Trenberth. "During Katrina, rainfall exceeded 12 inches near New Orleans."

It's not just warmer water on the surface that's powering the hurricanes; deeper warm water is too—at least in the Gulf of Mexico. Extending from the surface to a depth of 2,000 ft. or more is something scientists call the Loop Current, a U-shaped stream of warm water that flows from the Yucatán Straits to the Florida Straits and sometimes reaches as far north as the Mississippi River delta. Hurricanes that pass over the Loop typically get an energy boost, but the extra kick is brief, since they usually cross it and move on. But Rita and Katrina surfed it across the Gulf, picking up an even more powerful head of steam before slamming into the coastal states. Even if those unlucky beelines had been entirely random, the general trend toward warmer Gulf water may well have made the Loop even deadlier than usual.

"We don't know the temperature within the Loop Current," says Nan Walker, director of Louisiana State University's Earth Scan Laboratory. "It's possible that below the surface, it's warmer than normal. This needs to be investigated."

Other greenhouse-related variables may also be fueling the storms. Temperature-boosting carbon dioxide, for example, does not linger in the atmosphere forever. Some of it precipitates out in rain, settling partly on the oceans and sinking at least temporarily out of sight. But the violent frothing of the water caused by a hurricane can release some of that entrained CO_2, sending it back into the sky, where it resumes its role in the warming cycle. During Hurricane Felix in 1995, measurements taken in one area the storm struck showed local CO_2 levels spiking 100-fold.

So, are hurricanes actually speeding the effects of global warming and thus spawning even more violent storms? That's a matter of some dispute. While many scientists agree that this outgassing process goes on, not everyone agrees that it makes much of a difference. "The amount of CO_2 given off is fairly insignificant in terms of the total CO_2 in the atmosphere," says atmospheric scientist Chris Bretherton of the University of Washington in Seattle. "I am fairly confident in saying that there is no direct feedback from hurricanes."

Thus scientific uncertainty enters the debate—a debate already intensified by the political passions that surround any discussion of global warming. The fact is, there is plenty of room for doubt on both sides of the argument. Chris Landsea, a science and operations officer at the National Hurricane Center in Miami, is one of many experts who believe that global warming may be boosting the power of hurricanes—but only a bit, perhaps 1% to 5%. "A 100-mile-per-hour wind today would be a 105-mile-per-hour wind in a century," he says. "That is pretty tiny in comparison with the swings between hurricane cycles."

Scientific uncertainty enters the debate—a debate already intensified by the political passions that surround any discussion of global warming.

Skeptics are also troubled by what they see as a not inconsiderable bias in how hurricane researchers collect their data. Since most hurricanes spend the majority of their lives at sea—some never making land at all—it's impossible to measure rainfall precisely and therefore difficult to measure the true intensity of a storm.

What's more, historical studies of hurricanes like Emanuel's rely on measurements taken both before and during the era of satellites. Size up your storms in radically divergent ways, and you're likely to get radically divergent results. Even after satellites came into wide use—adding a significant measure of reliability to the data collected—the quality of the machines and the meteorologists who relied on them was often uneven. "The satellite technology available from 1970 to 1989 was not up to the job," says William Gray of Colorado State University. "And many people in non–U.S. areas were not trained well enough to determine the very fine differences between, say, the 130-m.p.h. wind speed of a Category 4 and, below that, a Category 3."

There's also some question as to whether there's a subtler, less scientific bias going on, one driven not by the raw power of the storms but by where they do their damage. Hurricanes that claw up empty coasts don't generate the same headlines as those that strike the places we like to live—and increasingly we like to live near the shore. The coastal population in the U.S. jumped 28% between 1980 and 2003. In Florida alone, the increase was a staggering 75%. Even the most objective scientists can be swayed when whole cities are being demolished by a hurricane.

"The storm activity this year is not necessarily higher than in previous high-activity years. It's just where they are going," says meteorologist Stan Goldenberg of the National Oceanic and Atmospheric Administration in Key Biscayne, Fla. "If you've got a guy shooting a machine gun but he's not shooting toward your neighborhood, it doesn't bother you."

Even correcting for our tendency to pay more attention to what is happening in our backyard, however, the global census of storms and the general measurement of their increasing power don't lie. And what those measurements tell scientists is that this already serious problem could grow a great deal worse—and do so very fast.

Some scientists are studying not just climate change but the even more alarming phenomenon of abrupt climate change. Complex systems like the atmosphere are known to move from one steady state to another with only very brief transitions in between. (Think of water, which when put over a flame becomes hotter and hotter until suddenly it turns into steam.) Ice cores taken from Greenland in the 1990s by geoscientist Richard Alley of Pennsylvania State University show that the last ice age came to an end not in the slow creep of geological time but in the quick pop of real time, with the entire planet abruptly warming in just three years.

"There are thresholds one crosses, and change runs a lot faster," Alley says. "Most of the time, climate responds as if it's being controlled by a dial, but occasionally it acts as if it's controlled by a switch." Adds Laurence Smith, an associate professor of geography at UCLA who has been studying fast climate change in the Arctic: "We face the possibility of abrupt changes that are economically and socially frightening."

Do we have the time to avert even a relatively slow climate change, or at least the nimbleness to survive it? That's what a lot of scientists are trying to determine. Japanese climatologists, for example, are using the Earth Simulator in Yokohama—one of the most powerful supercomputers in the world—to develop climate models that are more and more sophisticated. Scientists like geologist Claudia Mora of the University of Tennessee at Knoxville are going in another direction, studying isotopes locked in old tree rings to look for clues to past eras of heavy and light rainfall. Pair that information with global-temperature estimates for the same periods, and you can get a pretty good idea of how heat and hurricanes drive each other. "We've taken it back 100 years and didn't miss a storm," said Mora.

It's impossible to say whether any of that will convince the lingering global-warming skeptics. What does seem certain is that the ranks of those skeptics are growing thinner. In Washington successive administrations have ignored greenhouse warnings, piling up environmental debt the way we have been piling up fiscal debt. The problem is, when it comes to the atmosphere, there's no such thing as creative accounting. If we don't bring our climate ledgers back into balance, the climate will surely do it for us.

A Season a Tad off Color,
and Here's Why

By Avi Salzman
The New York Times, October 16, 2005

Fall arrives in New York state like a prom queen, draped in boastful reds, yellows and rusty browns, perfumed with wood smoke. It saunters through the Appalachians and down the Catskills before taking its final bow in Westchester and New York City.

The season is important here not only for the spectacle it creates, but also for the tourist economy it supports and the cultural symbol it represents; no season is more strongly identified with the Northeast than fall. Even 150 years ago, Henry David Thoreau realized how special it was.

"October is the month of painted leaves," Thoreau wrote in his 1862 essay "Autumnal Tints." "Their rich glow now flashes round the world."

But what if the glow wasn't so rich?

That is a question some scientists have already begun to ask, and they don't think it is hypothetical. Already this year, throughout the Northeast, the foliage has been late, and, some say, unspectacular. Scientists at the University of New Hampshire project that shifts in the climate caused by global warming will progressively dull the leaves throughout southern New England and New York over the next century. Maples will move north and the remaining oaks and hickories will change colors later and with less verve, they say. If the projections are correct, leaf-peepers may be forced as far north as Canada to for their fix of foliage.

Indeed, one scientist says autumn in the region has already lost its brilliance.

"We haven't had a really great display in the last 10 years," said Barrett Rock, a professor in natural resources and a researcher at the Complex Systems Research Center at the University of New Hampshire who has studied the effects of global warming on the autumn landscape from New York to Maine. Dr. Rock was the lead author on a federally financed 2001 report, the New England Regional Assessment, that assessed the effects of climate change on the Northeast.

This year, the problem has been intense heat in late summer and early fall, which delayed the changing of the leaves and, Dr. Rock said, dulls the foliage. The average temperature in New York in August was 70.6 degrees, nearly 4 degrees higher than the average since 1895, and the fifth highest August average in that period, according to the National Oceanic and Atmospheric Administration. The heat did not let up in September, when it was again 4 degrees warmer than average.

But global warming is a gradual phenomenon, and one warm year does not mean that fall will be this warm every year.

Nonetheless, scientists like Dr. Rock say there is ample reason to worry.

In the 2001 report, Dr. Rock and fellow researchers at the University of New Hampshire and other universities in the region found that New York's average temperature had increased by 1 degree over the last century. The scientists used two different models to project what the Northeast would look like over the next 100 years.

Global warming is a gradual phenomenon, and one warm year does not mean that fall will be this warm every year.

Both models, Dr. Rock noted in a telephone interview, are considered "middle of the road," predicting gradual but not extreme warming trends. But both project that global warming will accelerate dramatically in the region. One forecasts an overall temperature increase of 6 degrees by 2090. That would leave Boston looking and feeling like Richmond, Va. The other model projected a climate like Atlanta's, a full 10 degrees warmer than Boston's, according to the report.

This year another University of New Hampshire environmental scientist, Cameron P. Wake, issued a report in conjunction with Clean Air Cool Planet, a nonprofit organization that studies climate change in the Northeast. Using different methods from the earlier study, the new report confirmed the warming trend in the Northeast throughout the 20th century, and showed it increasing over the last 30 years. That report did not focus on foliage but found other significant effects to plants and trees, including progressively later bloom dates for lilac trees.

The warming trends affect foliage in various ways, Dr. Rock said. As greenhouse gases accumulate and the earth warms, summer weather will linger into fall. That will delay the frosts that degrade chlorophyll and trigger leaves to change. It takes at least one frost in which temperatures fall to 30 degrees Fahrenheit or below to start the changes in the leaves. If those fall frosts are delayed, or the chlorophyll breaks down more gradually, leaf-peepers won't see the vibrant reds they have come to expect, Dr. Rock said.

Last week, Marian Travis was decorating a lamp post near her house in Ossining, N.Y., with colorful fake leaves when she realized that something felt wrong—though she was getting her neighborhood ready for fall, it felt more like summer.

"It didn't even feel like fall," she said. "It is fall by the calendar, so you have to put the decorations up."

Ms. Travis said she thinks fall has changed since she was a girl. "I remember in the fall, I'd walk home from school and there were so many leaves in the street and the gutter," she said. "They don't have them anymore. I don't think the fall is like what we used to have way back."

Global warming will affect which tree species can survive in the state and which will move north to cooler climates, scientists say.

Maples, with their stunning red, yellow and orange leaves, are beauties in the fall. Thoreau positively gushed over them, writing that they seem to absorb "all the sunny warmth of the season."

Global warming will affect which tree species can survive in [New York State] and which will move north to cooler climates, scientists say.

New York is full of maples, too, but changes in temperature, air quality and rainfall, the 2001 report said, will change the mix of trees in the state's forests.

Birches and beeches, two other species known for their foliage, can also be expected to migrate north, the report noted.

It details other possible effects of global warming on foliage, including increases in plant diseases and insect outbreaks. Warming also increases ground-level ozone, which can damage leaves.

But it is not just scientists who have a stake in the changing autumn landscape. The colors of fall are big business for New York's tourism industry. The state maintains a Web site (www.iloveny.com/fall/pages/foliagereport.html) that specifically tracks the progression of the foliage, enlisting about 55 volunteers from all over the state who call and write in to report the quality of the leaves in their areas. A map highlights the regions of the state where the foliage is brightest.

Leaf-peepers make up an important portion of New York's tourist population, said Charles Gargano, chairman of the Empire State Development Corporation, which runs the state's I Love New York tourism Web site (www.iloveny.state.ny.us/). Summer is the king of tourist seasons in the state, but fall still attracts 21 percent of the yearly visitors, he said.

"Certainly, we depend on beautiful foliage," he said. "A lot of people travel around the state to see the foliage."

Peter Barton, a member of the family that owns Barton Orchard, a "pick-your-own" farm in Poughquag that attracts more than 100,000 people in the fall, said the last decade has been marked by more extreme weather, another expected effect of global warming.

"From my point of view as a grower, the last decade we've seen much more volatility," he said. "The storms have been more volatile."

He said he was concerned about the effect of global warming on plants, even if the increase in temperature isn't as much as 6 or 10 degrees.

"A simple 1 or 2 degree change can change the organisms that can exist here," he said.

If that happens, and the autumn prom queen starts dressing in brown or dull green, autumn in New York could look a lot different.

Arid Arizona Points to Global Warming as Culprit

By Juliet Eilperin
The Washington Post, February 6, 2005

Reese Woodling remembers the mornings when he would walk the grounds of his ranch and come back with his clothes soaked with dew, moisture that fostered enough grass to feed 500 cows and their calves.

But by 1993, he says, the dew was disappearing around Cascabel—his 2,700-acre ranch in the Malpai borderlands straddling New Mexico and Arizona—and shrubs were taking over the grassland. Five years later Woodling had sold off half his cows, and by 2004 he abandoned the ranch.

"How do you respond when the grass is dying? You hope to hell it starts to rain next year," he says.

When the rain stopped coming in the 1990s, he and other Southwest ranchers began to suspect there was a larger weather pattern afoot. "People started talking about how we've got some major problems out here," he said in an interview. "Do I believe in global warming? Absolutely."

Dramatic weather changes in the West—whether it is Arizona's decade-long drought or this winter's torrential rains in Southern California—have pushed some former skeptics to reevaluate their views on climate change. A number of scientists, and some Westerners, are now convinced that global warming is the best explanation for the higher temperatures, rapid precipitation shifts, and accelerated blooming and breeding patterns that are changing the Southwest, one of the nation's most vulnerable ecosystems.

In the face of shrinking water reservoirs, massive forest fires and temperature-related disease outbreaks, several said they now believe that warming is transforming their daily lives. Although it has rained some during the past three months, the state is still struggling with a persistent drought that has hurt its economy, costing cattle-related industries $2.8 billion in 2002.

"Everyone's from Missouri: When they see it, they believe it," said Gregg Garfin, who has assessed the Southwest's climate for the federal government since 1998. "When we used to talk about climate, eyes would glaze over. . . . Then the drought came. The phone started ringing off the hook."

Jonathan Overpeck, who directs the university- and government-funded Institute for the Study of Planet Earth at the University of Arizona, said current drought and weather disruptions signal what is to come over the next century. Twenty-five years ago, he said, scientists produced computer models of the drought that Arizona is now experiencing.

"It's going to get warmer, we're going to have more people, and we're going to have more droughts more frequently and in harsher terms," Overpeck said. "We should be at the forefront of demanding action on global warming because we're at the forefront of the impacts of global warming. . . . In the West we're seeing what's happening now."

There are dissenters who say it is impossible to attribute the recent drought and higher temperatures to global warming. Sherwood Idso, president of the Tempe, Ariz.–based Center for the Study of Carbon Dioxide and Global Change, said he does not believe the state's drought "has anything to do with CO_2 or global warming," because the region experienced more-severe droughts between 1600 and 1800. Idso, who also said he did not believe there is a link between human-generated carbon dioxide emissions and climate change, declined to say who funds his center.

The stakes are enormous for Arizona, which is growing six times faster than the national average and must meet mounting demands for water and space with scarce resources. Gov. Janet Napolitano (D) is urging Arizonans to embrace "a culture of conservation" with water, but some conservationists and scientists wonder whether that will be enough.

Dale Turner of the Nature Conservancy tracks changes in the state's mountaintop "sky islands"—a region east and south of Tucson that hosts a bevy of rare plants and animals. Human activities over the past century have degraded local habitats, Turner said, and now climate change threatens to push these populations "over the edge."

The Mount Graham red squirrel, on the federal endangered species list since 1987, has been at the center of a long-running fight between environmentalists and development-minded Arizonans. Forest fires and rising temperatures have worsened the animals' plight as they depend on Douglas firs at the top of a 10,720-foot mountain for food and nest-building materials. The population has dipped from about 562 animals in spring 1999 to 264 last fall.

"They are so on the downhill slide," said Thetis Gamberg, a U.S. Fish and Wildlife biologist who has an image of the endangered squirrel on her business card.

Atop Mount Graham, the squirrels' predicament is readily visible. Mixed conifers are replacing Douglas firs at higher altitudes, and recent fires have destroyed other parts of the forest, depriving the animals of the cones they need.

Environmentalists such as Turner worry about the disappearance of the Mount Graham squirrel, the long-tailed, mouselike vole and native wet meadows known as cienegas, but many lawmakers and state officials are more focused on the practical question of water supply.

Arizona gets its water from groundwater and rivers such as the massive Colorado, a 1,450-mile waterway that supplies water to seven states: Arizona, California, Colorado, Nevada, New Mexico, Utah and Wyoming.

"Weather is like rolling the dice, and climate change is like loading the dice."—C. Mark Eakin, the National Oceanic and Atmospheric Administration

The recent drought and changing weather patterns have shrunk the western snowpack and drained the region's two biggest reservoirs, lakes Mead and Powell, to half their capacity. More precipitation is falling as rain instead of snow, and it is coming earlier in the year, which leads to rapid runoff that disappears quickly.

Scientists at Scripps Institution of Oceanography predict that by 2090 global warming will reduce the Sierra Nevada snowpack, which accounts for half of California's water reserves, by 30 percent to 90 percent.

"It makes water management more challenging," said Kathy Jacobs, who spent two decades managing state water resources before joining the University of Arizona's Water Resources Research Center. "You can either reduce demand or increase supply."

Water managers have just begun to consider climate change in their long-term planning. Forest managers have also started asking for climate briefings, now that scientists have documented that short, wet periods followed by drought lead to the kind of giant forest fires that have been devastating the West.

This month, scientists at the National Center for Atmospheric Research in Boulder, Colo., published a study showing that worldwide, regions suffering from serious drought more than doubled in area from the early 1970s to the early 2000s, with much of the change attributed to global warming. A separate recent report in the journal *Science* concluded that higher temperatures could cause serious long-term drought over western North America.

C. Mark Eakin, a paleoclimatologist at the National Oceanic and Atmospheric Administration who co-wrote the study in *Science*, said historical climate records suggest the current drought could just be the beginning.

"When you've got an increased tendency toward drought in a region that's already stressed, then you're just looking for trouble," Eakin said. "Weather is like rolling the dice, and climate change is like loading the dice."

Still, Arizona politicians remain divided on how to address global warming. Sen. John McCain (R-Ariz.) has led the national fight to impose mandatory limits on industrial carbon dioxide emissions that are linked to warming, though his bill remains stalled.

"We'll win on this issue because the evidence continues to accumulate," McCain said in an interview. "The question is how much damage will be done until we do prevail."

But other Arizona Republicans are resistant. State Sen. Robert Blendu, who opposed a bill last year to establish a climate change study committee, said he wants to make sure politicians "avoid the public knee-jerk reaction before we get sound science."

That mind-set frustrates ranchers such as Woodling, who is raising 10 grass-fed cows on a leased pasture. At age 69, he will never be able to rebuild his herd, he said, but he believes politicians have an obligation to help restore the environment.

"Man has been a great cause of this, and man needs to address it," he said.

Climate Change Means Big Changes in Puget Sound

By Craig Welch
The Seattle Times, October 18, 2005

Rising global temperatures are taking their toll on Puget Sound, as less mountain snow funnels less freshwater into estuaries, rising oceans transform salt marshes, and changes to the food web alter life for everything from lingcod to orcas, a new study says.

The report released yesterday, the first serious attempt by University of Washington scientists and the state to gauge the impact of global warming on Puget Sound, suggests that climate change will continue to echo across the ecosystem, upsetting links between plants and animals and complicating efforts to manage the threat of a growing human population.

"It's not like we're going to wake up tomorrow and everything will be dead," said Jan Newton, a UW oceanographer who contributed to the 35-page report by the Puget Sound Action Team, a state agency that monitors the Sound's health.

"But we also know that when organisms experience catastrophe, it's most often because they're assaulted by more than one problem at a time. The sooner we recognize that things are under pressure because of climate change, we can look at the stressors we can do something about."

The state has issued biannual reports in recent years on the health of Puget Sound. Until yesterday, however, no one had created a cohesive picture from all the latest climate science.

"It's frightening and baffling," said Brad Ack, executive director of the Puget Sound Action Team. "I was surprised by how much we've already experienced—some of the most significant change in North America. What's less certain is how it will all play out."

Already, the region's signature body of water is warming faster and its water levels are rising quicker than waters in many other parts of the world, the report says.

While average global temperatures rose 1.1 degrees Fahrenheit during the 20th century, Northwest winters have warmed 2.7 degrees since 1950, in part because of cycles in ocean conditions. Scientists don't know how much of that to attribute to global warming.

Meanwhile, water levels in south Puget Sound are expected to increase 1.3 feet by 2050, higher than in most areas of the world because changes in Pacific Ocean wind patterns drive more seawater into Puget Sound. The change will be compounded because geological factors are causing land in the Olympia and Tacoma areas to sink about an inch every 12 years.

Runoff from the 10,000 rivers and streams that spill into the Sound is already shifting, according to the report: About 13 percent less freshwater flows to Puget Sound now than in 1948, and snowmelt is coming an average of 12 days earlier.

With lower snowfall, more of the Northwest's precipitation comes as rain, the study says, so flooding is likely to increase because the water isn't held in mountain snowpack. And that could affect how sediment and debris is washed into the Sound.

Such simple-seeming swings can have wide impacts.

Rising water increases erosion, and also threatens the habitat of kelp and other grassy plants that incubate dozens of species of fish and vegetation that can't survive without it.

Geological factors are causing land in the Olympia and Tacoma areas to sink about an inch every 12 years.

Warming water could increase harmful algae blooms, which can contaminate shellfish. When those blooms die, they suck oxygen out of the water, causing "dead zones" like an area in Hood Canal where thousands of fish have been killed.

Lower water levels in rivers and streams already have hurt salmon runs. But when the largest volumes of freshwater come earlier in the year, it can compound problems for many other creatures in the Sound, said Nate Mantua, a professor of atmospheric sciences at UW's Joint Institute for the Study of the Atmosphere and Oceans.

Rivers flush plankton and other nutrients into the Sound, as well as helping to cool estuaries and moderate their salt levels. That helps determine how much food wells up from the cold, dark bottom.

But as the volume and timing of those river flows change, the tiny creatures that eat the plankton may still emerge when they always have, which means they may miss their primary food source.

In turn, that means those tiny creatures may not be available to be eaten when herring or other bottom fish spawn, or when shorebirds arrive during migration.

What's worrisome to the scientists is that even if one or two of those factors is altered, the whole food chain would have to adjust, potentially changing the diversity of fish and mammals in the Sound.

"Usually what we find is that rapid change favors the generalist; if you rip out your lawn, what repopulates it first . . . dandelions, right?" said Phil Mote, a climatologist with UW's Climate Impacts

Group who participated in the study. "So stress could push some more specialized species over the brink and make it easier for species that can live anywhere, many of which are invasive species."

But all of the complicated functions are still poorly understood.

"It's like trying to predict the economy," said Newton, the UW oceanographer. "We have tools and models, but there's just a lot we don't know yet."

For example, eelgrass, an important nursery for fish, may actually benefit from changes in freshwater flows. And when plankton blooms come at different times, many species lose, but others win.

"Climate change is shifting that whole balance," said Amy Snover, a research scientist with UW's Climate Impacts Group. "What happened in the past is no longer going to be a way to judge what will happen in the future."

Even as scientists seek to better predict the future, Ack said politicians and governments should consider broader implications when managing everything from shorelines to water supplies, salmon and hydropower.

"I'm committed that every time we have a discussion about these issues, we're asking how climate change fits in," he said.

A Darkening Sky

BY CHARLES W. PETIT
U.S. NEWS & WORLD REPORT, MARCH 17, 2003

V. "Ram" Ramanathan sat on an airliner heading south from Bombay. Ahead were the Maldives, an archipelago near the equator, where the atmospheric scientist from the Scripps Institution of Oceanography near San Diego planned to set up instruments to study haze and weather. He expected that results from the international project would come slowly and be of interest only to specialists. He was not prepared for what he saw just gazing out the plane window.

As he took off from Bombay, the layers of brown gunk in the sky were no surprise. Pollution controls on factories and vehicles are rare in his native land. Hundreds of millions of its citizens burn low-quality coal, wood, and cow dung for cooking and heating. But nearly 1,000 miles later over the open sea, the dirty pall still had not given way to blue sky and white clouds. "The haze just kept going and going. It didn't even seem to thin out. I was thinking, this is something big."

It is. Since Ramanathan's 1998 flight, scientists have realized that the pall he saw is just part of a vast brown cloud that often extends thousands of miles east, across China. A stew of dust, ash, and smoke from fires and industry, the cloud threatens the health of the billions who live under it. The fine particles, or aerosols, also warm some areas and cool others, drying up storm clouds and perhaps even shifting India's life-giving monsoon. In many places the haze swamps greenhouse gases as a climate-changing force, say scientists. The atmospheric havoc in Asia may even play a role in El Nino, the climate cycle now drenching the southern United States.

Much of this picture is still fuzzy, but scientists are working to sharpen it. Ramanathan and his Scripps colleague Paul Crutzen, a chemist and Nobel laureate, made a start with their Indian Ocean Experiment in the late 1990s, which studied haze from a score of ground stations and from aircraft. Their glimpses of the cloud's extent and impacts helped set off an explosion of similar studies across India, off Japan and Korea, and in China, which has launched the largest single scientific project in the country's history to analyze aerosols and climate. And it has spawned a new United

Nations effort called Project Asian Brown Cloud. Led by Ramanathan and Crutzen, it is organizing a massive study of the pollution's sources and effects, and what to do about it.

In a way, Asia with its dirty, fast-growing industry is repeating on a far vaster scale the smoky evolution of European and U.S. industry in the 19th and early 20th centuries. Coal consumption in China, for example, was 50 percent higher than in the United States in 1999 and could be twice as high by 2010. Across Asia, coal heats houses and cooks meals. Smoke from agricultural burning and wildfires adds to the brew. In China, the haze sometimes starts as dust blowing off western deserts, "but it picks up all kinds of toxic pollutants as it travels," says F. Sherwood Rowland, a University of California–Irvine chemist who received a Nobel Prize for work on ozone. "We can detect Asian aerosols blowing all the way across the U.S."

Yet just five years ago, Ramanathan could be startled by the pall he saw from the plane window because experts still thought of smog outbreaks as local, covering a city or filling a river valley. Until recently nobody had seen the goop all in one glance. Cameras on early weather satellites were calibrated for clouds but not hazes. But new full-color satellite camera systems now send images of a nearly continuous, 2-mile-thick blanket of sulfates, soot, organic compounds, dust, fly ash, and other crud draped across much of India, Bangladesh, and Southeast Asia, including the industrial heart of China.

The sand-colored air of Los Angeles is pristine by comparison. When Chinese scientists told U.S. colleagues about foul air back home, "we'd say we have smog here too," says Lorraine Remer, who analyzes satellite data at NASA's Goddard Space Flight Center. "Then we saw the extinction numbers"—satellite data on how much the brown cloud dims light. Across much of Asia, they were several times higher than anything ever seen in American smog. "We were standing there not believing it," she says. In and around India, the researchers found sunlight was reduced by 10 percent. Crop scientists say this is enough to reduce rice yields by 3 percent to 10 percent across much of the country.

Ground data in China show the same thing. In Beijing, airborne particulates are routinely five times as high as in Los Angeles. Donald Blake, an atmospheric chemist at the University of California–Irvine, says that a colleague on a visit asked a group of kindergartners to draw the sky. They all reached for the gray crayon.

It's worse than unsightly. India has 23 cities of more than 1 million people; not one meets World Health Organization pollution standards. Indoor smoke from poorly vented fires is blamed for half a million premature deaths annually in India alone, mostly women and children. In southern China and Southeast Asia, as many as 1.4 million people die annually from pollution-related respiratory ills.

Disturbing Effect

Researchers are coming to realize that, through a long chain of effects, the brown cloud may also be to blame for drought and flooding. Scientists' understanding of how aerosols shape climate is not nearly as well developed as it is for greenhouse gases like carbon dioxide, still No. 1 on any list of human impacts on climate. "But one common aspect," says Ramanathan, "is that the haze and its heating of the atmosphere is sufficient to disturb climate a lot."

Unlike the whitish sulfate particles from cleaner-burning power plants in the United States and Europe, the Asian hazes are dark with soot. As a result, they absorb sunlight and can double the rate at which it warms the atmosphere several thousand feet up, while shading and cooling the ground below. Some scientists think that the net effect is to boost global warming. But the more certain impact of the hazes is on rainfall, says Jeff Kiehl of the National Center for Atmospheric Research in Boulder, Colo. "They are radically changing the temperature profile of the atmosphere in many areas, with a big impact on where rain falls and how much."

[Asian hazes] absorb sunlight and can double the rate at which it warms the atmosphere several thousand feet up, while shading and cooling the ground below.

By cooling the northern Indian Ocean, the haze reduces evaporation, cutting the water supply for rainfall. On land, the warm air aloft acts as a lid on cloud formation, quashing the convection that feeds thunderstorms. And the aerosols themselves seed the formation of tiny mist particles—so many that they suck water out of the air and choke off the growth of larger drops that would fall as rain. While the haze particles dry out the land, the rain does fall over the sea, where larger, natural sea-salt particles promote droplet growth. "We're shifting rain from the land to the ocean," says Daniel Rosenfeld of the Hebrew University of Jerusalem.

At least that's the theory, and there are signs it may be happening. Some computer climate models predict that the hazes over India should displace the annual monsoon rains, leading to floods in the south and east of the country while drying the north and shrinking the vital Himalayan snowpack. "That's just the pattern we are starting to see emerge," says Surabi Menon of the Goddard Institute for Space Studies in New York City.

In southeastern China, where haze has cut sunlight by 2 percent to 3 percent every 10 years since the 1950s, temperatures are dropping, while rising elsewhere in the country, presumably because of greenhouse gases. The changed temperature patterns have rerouted storm tracks, one recent Chinese study said. The study blamed the shift for severe floods in the nation's south in recent years, coupled with drought in the north. It ranked the new weather pattern as the greatest sustained change in China's climate in more than 1,000 years.

Some scientists also suspect that the pollution cloud could be cooling the sea surface and slowing evaporation in the far western Pacific, off Asia. The effects could ripple across half the globe to the United States, because the western Pacific is the breeding ground for El Niños, the bouts of Pacific warming that change rainfall across the Americas and beyond.

All of this is enough to make Asia's brown cloud, and the sparser hazes elsewhere, into a global climate threat. Fortunately, hazes are far easier to counter than greenhouse gases like carbon dioxide. Clean up industry and smother the fires, and in a few weeks rain would wash the skies clean. Carbon dioxide, in contrast, lingers for centuries, and ordinary pollution controls can't touch it.

Going After Hazes

Some scientists, distressed at the reluctance of the U.S. government and many developing nations to tackle greenhouse gases, hope that the relatively easier task of curbing fine particles could kick-start international efforts to address climate change. Going after hazes, particularly those heavy with soot, is "a no-lose situation as far as I'm concerned," says Stanford University atmospheric researcher Mark Jacobson.

The Chinese government, rattled by the data on the country's polluted air, is doing just that. For both health and weather reasons, it has largely replaced home use of coal with cleaner-burning natural gas in big cities and is starting to require catalytic converters on vehicles. China also hopes to restore blue skies to Beijing in time for the 2008 Olympics.

At the same time, some scientists worry that major assaults on aerosols might divert attention from the far tougher problem of carbon dioxide and other culprits in global warming. Jacobson laments that President Bush cited the climate impact of soot as one reason to abandon the Kyoto climate change agreement, which deals with greenhouse gases. "You can't stop with aerosols," Jacobson says. "You definitely have to go after the greenhouse gases, too."

But the lesson of the Asian brown cloud, says Ramanathan, is that there's more to global change than greenhouse warming. "If all you deal with is CO_2, then you don't understand climate at all."

III. Armageddon Approaching? What the Future Holds

Editor's Introduction

Given the complexity of the Earth's climate, scientists are not certain precisely what global warming has in store for the years ahead. This may seem counterintuitive: One would expect global warming to cause exactly that—a fairly steady increase in the planet's temperature. However, many experts believe that the warming currently under way may eventually flip a climatic switch of sorts, setting in motion abrupt, unpredictable, and severe climate change. The entries in this chapter, "Armageddon Approaching? What the Future Holds," consider how and to what degree global warming may impact the future.

Richard B. Alley, in the first selection, "Abrupt Climate Change," compares the current state of the climate to a canoe, and human beings, through fossil-fuel consumption, are effectively shaking that canoe. At some point, Alley hypothesizes, the canoe will be pushed to a particular threshold, at which point it will abruptly capsize. Alley explores a number of possible scenarios that could transpire if climate change is not forestalled, among them a slowing of the North Atlantic conveyor, which could cause Europe to become significantly cooler. In "Crossing the Threshold," Peter Bunyard offers a similar analysis, taking into account the myriad variables that could come into play when this environmental tipping point is reached.

Climate change is expected to endanger much of the planet's wildlife. As Jane Kay notes in "Dire Warming Warning for Earth's Species," scientists predict that as many as one in four of the planet's species could be extinct or en route to extinction by the year 2050 if global-warming trends continue.

A recent study described by Michael Powell in the next entry, "Northeast Seen Getting Balmier," suggests that climate change will cause the northeastern United States to become warmer, resulting in shorter winters and forcing the northward migration of such native species as the maple tree. In addition, the hotter temperatures could threaten the region's water resources and increase instances of heat-related illness and death. In the western United States, global warming could reduce snowpack in the Sierra Mountains by as much as 89 percent, Miguel Bustillo reports in "Risk to State Dire in Climate Study." Such a scenario would imperil the water supply throughout California and threaten the state's wine harvest, among other things.

On a per-capita basis, Africa's greenhouse gas emissions constitute a mere fraction of those produced by the other inhabited continents. Yet, as Scott Fields explains in the final piece in this chapter, "Continental Divide: Why Africa's Climate Change Burden Is Greater," global warming will likely exact a higher toll from Africa in human and ecological terms than from any other continent. Economically dependent on subsistence agriculture and burdened by endemic poverty, political instability, and the HIV/AIDS crisis, the African populace will likely find it difficult to adapt to the changes wrought by global warming, the signs of which are already emerging. Consequently, experts expect continued climate change to negatively impact the continent's food and water supplies, leading perhaps to military conflict over dwindling resources, as well as widespread famine and disease.

Abrupt Climate Change

By Richard B. Alley
Scientific American, November 2004

In the Hollywood disaster thriller *The Day after Tomorrow*, a climate catastrophe of ice age proportions catches the world unprepared. Millions of North Americans flee to sunny Mexico as wolves stalk the last few people huddled in freeze-dried New York City. Tornadoes ravage California. Giant hailstones pound Tokyo.

Are overwhelmingly abrupt climate changes likely to happen anytime soon, or did Fox Studios exaggerate wildly? The answer to both questions appears to be yes. Most climate experts agree that we need not fear a full-fledged ice age in the coming decades. But sudden, dramatic climate changes have struck many times in the past, and they could happen again. In fact, they are probably inevitable.

Inevitable, too, are the potential challenges to humanity. Unexpected warm spells may make certain regions more hospitable, but they could magnify sweltering conditions elsewhere. Cold snaps could make winters numbingly harsh and clog key navigation routes with ice. Severe droughts could render once fertile land agriculturally barren. These consequences would be particularly tough to bear because climate changes that occur suddenly often persist for centuries or even thousands of years. Indeed, the collapses of some ancient societies—once attributed to social, economic and political forces—are now being blamed largely on rapid shifts in climate.

The specter of abrupt climate change has attracted serious scientific investigation for more than a decade, but it has only recently captured the interest of filmmakers, economists and policymakers. Along with more attention comes increasing confusion about what triggers such change and what the outcomes will be. Casual observers might suppose that quick switches would dwarf any effects of human-induced global warming, which has been occurring gradually. But new evidence indicates that global warming should be more of a worry than ever: it could actually be pushing the earth's climate faster toward sudden shifts.

Jumping Back and Forth

Scientists might never have fully appreciated the climate's ability to lurch into a radically different state if not for ice cores extracted from Greenland's massive ice sheet in the early 1990s. These colossal rods of ice—some three kilometers long—entomb a remarkably clear set of climate records spanning the past 110,000 years. Investigators can distinguish annual layers in the ice cores and date them using a variety of methods; the composition of the ice itself reveals the temperature at which it formed.

Such work has revealed a long history of wild fluctuations in climate—long deep freezes alternating with brief warm spells. Central Greenland experienced cold snaps as extreme as six degrees Celsius in just a few years. On the other hand it achieved roughly half of the heating sustained since the peak of the last ice age—more than 10 degrees degrees C—in a mere decade. That jump, which occurred about 11,500 years ago, is the equivalent of Minneapolis or Moscow acquiring the relatively sultry conditions of Atlanta or Madrid.

Not only did the ice cores reveal what happened in Greenland, but they also hinted at the situation in the rest of the world. Researchers had hypothesized that the 10-degree warming in the north was part of a warming episode across a broad swath of the Northern Hemisphere and that this episode enhanced precipitation in that region and far beyond. In Greenland itself, the thickness of the annual ice layers showed that, indeed, snowfall had doubled in a single year. Analyses of old air bubbles caught in the ice corroborated the prediction of increased wetness in other areas. In particular, measurements of methane in the bubbles indicated that this swamp gas was entering the atmosphere 50 percent faster during the intense warming than it had previously. The methane most likely entered the atmosphere as wetlands flooded in the tropics and thawed up north.

The cores also contained evidence that helped scientists fill in other details about the environment at various times. Ice layers that trapped dust from Asia indicated the source of prevailing winds, for instance. Investigators concluded that the winds must have been calmer during warm times because less windblown sea salt and ash from faraway volcanoes accumulated in the ice. And the list of clues goes on [see "Greenland Ice Cores: Frozen in Time," by Richard B. Alley and Michael L. Bender; *Scientific American*, February 1998].

Intense, abrupt warming episodes appeared more than 20 times in the Greenland ice records. Within several hundreds or thousands of years after the start of a typical warm period, the climate reverted to slow cooling followed by quick cooling over as short a time as a century. Then the pattern began again with another warming that might take only a few years. During the most extreme cold conditions, icebergs strayed as far south as the coast of Portugal. Lesser challenges probably drove the Vikings out of Greenland during the most recent cold snap, called the Little Ice Age, which started around A.D. 1400 and lasted 500 years.

Sharp warm-ups and cool-downs in the north unfolded differently elsewhere in the world, even though they may have shared a common cause. Cold, wet times in Greenland correlate with particularly cold, dry, windy conditions in Europe and North America; they also coincide with anomalously warm weather in the South Atlantic and Antarctica. Investigators pieced together these regional histories from additional clues they found in the ice of high mountain glaciers, the thickness of tree rings, and the types of pollen and shells preserved in ancient mud at the bottoms of lakes and oceans, among other sources.

The evidence also revealed that abrupt shifts in rainfall have offered up challenges rivaling those produced by temperature swings. Cold times in the north typically brought drought to Saharan Africa and India. About 5,000 years ago a sudden drying converted the Sahara from a green landscape dotted with lakes to the scorching, sandy desert it is today. Two centuries of dryness about 1,100 years ago apparently contributed to the end of classic Mayan civilization in Mexico and elsewhere in Central America. In modern times, the El Niño phenomenon and other anomalies in the North Pacific occasionally have steered weather patterns far enough to trigger surprise droughts, such as the one responsible for the U.S. dust bowl of the 1930s.

Threshold crossings caused history's most extreme climate flips—and point to areas of particular concern for the future.

Point of No Return

Be they warm spells, cold snaps or prolonged droughts, the precipitous climate changes of the past all happened for essentially the same reason. In each case, a gradual change in temperature or other physical condition pushed a key driver of climate toward an invisible threshold. At the point that threshold was crossed, the climate driver—and thus the climate as well—flipped to a new and different state and usually stayed there for a long time.

Crossing a climate threshold is similar to flipping a canoe. If you are sitting in a canoe on a lake and you lean gradually to one side, the canoe tips, too. You are pushing the canoe toward a threshold—the position after which the boat can no longer stay upright. Lean a bit too far, and the canoe overturns.

Threshold crossings caused history's most extreme climate flips—and point to areas of particular concern for the future. To explain the icy spells recorded in Greenland's ice cores, for example, most scientists implicate altered behavior of currents in the North Atlantic, which are a dominant factor in that region's long-term weather patterns.

Eastern North America and Europe enjoy temperate conditions (like today's) when salty Atlantic waters warmed by southern sunshine flow northward across the equator. During far northern winters, the salty water arriving from the south becomes cold and dense enough to sink east and west of Greenland, after which it migrates southward along the seafloor. Meanwhile, as the cooled water sinks, warm currents from the south flow northward to take its place. The sinking water thereby drives what is called a conveyor belt circulation that warms the north and cools the south.

> *Human-induced increases in atmospheric concentrations of greenhouse gases . . . are promoting global warming.*

Ice cores contain evidence that sudden cold periods occurred after the North Atlantic became less salty, perhaps because freshwater lakes burst through the walls of glaciers and found their way to the sea. Researchers identify this rush of freshwater as the first phase of a critical threshold crossing because they know freshening the North Atlantic can slow or shut off the conveyor, shifting climate as a result.

Diluted by water from the land, seawater flowing in from the south would become less salty and thus less dense, possibly to the point that it could freeze into sea ice before it had a chance to sink. With sinking stopped and the conveyor halted, rain and snow that fell in the north could not be swept into the deep ocean and carried away. Instead they would accumulate on the sea surface and freshen the North Atlantic even more. The conveyor then would stay quiet, leaving nearby continents with climates more like Siberia's.

Chilling Warmth

Eight thousand years have passed since the last of the biggest North Atlantic cold snaps. Could it be that humans are actually "leaning" in the right direction to avoid flipping the climate's canoe? Perhaps, but most climate experts suspect instead that we are rocking the boat—by changing so many aspects of our world so rapidly. Particularly worrisome are human-induced increases in atmospheric concentrations of greenhouse gases, which are promoting global warming [see "Defusing the Global Warming Time Bomb," by James Hansen; *Scientific American*, March; www.sciam.com/ontheweb].

The United Nations–sanctioned Intergovernmental Panel on Climate Change has predicted that average global temperatures will rise 1.5 to 4.5 degrees C in the next 100 years. Many computer models that agree with this assessment also predict a slowdown of the North Atlantic conveyor. (As ironic as it may sound, gradual warming could lead to a sudden cooling of many degrees.) Uncertainties abound, and although a new ice age is not thought credible, the resulting changes could be notably larger than they were during the Little Ice Age, when the Thames in London froze and glaciers rumbled down the Alps.

Perhaps of greater concern than cold spells up north are the adverse effects that would probably strike other parts of the world concurrently. Records of climate across broad areas of Africa and Asia that typically benefit from a season of heavy monsoons indicate that these areas were particularly dry whenever the North Atlantic region was colder than the lands around it. Even the cooling from a conveyor slowdown might be enough to produce the drying. With billions of people relying on monsoons to water crops, even a minor drought could lead to widespread famine.

The consequences of future North Atlantic freshening and cooling may make life more difficult even for people living in regions outside the extreme cold or drought. Unease over such broad impacts spurred the U.S. Department of Defense to request that a think tank called the Global Business Network assess the possible national security implications of a total shutdown of the North Atlantic conveyor. Many scientists, including me, think that a moderate slowdown is much more likely than a total shutdown; either way, the seriousness of the potential outcome makes considering the worst-case implications worthwhile. As the Global Business Network report states, "Tensions could mount around the world. . . . Nations with the resources to do so may build virtual fortresses around their countries, preserving resources for themselves. Less fortunate nations . . . may initiate in struggles for access to food, clean water, or energy."

Floods and Droughts

Even if a slowdown of the North Atlantic conveyor never happens, global warming could bring about troubling threshold crossings elsewhere. The grain belts that stretch across the interiors of many midlatitude continents face a regional risk of prolonged drought. Most climate models produce greater summertime drying over these areas as average global temperatures rise, regardless of what happens in the North Atlantic. The same forecasts suggest that greenhouse-induced warming will increase rainfall overall, possibly in the form of more severe storms and flooding; however, those events—significant problems on their own—are not expected to offset the droughts.

Summer drying could cause a relatively mild drought to worsen and persist for decades or more. This transition would occur because of a vulnerability of the grain belts: for precipitation, they rely heavily on rainfall that local plants recycle rather than on new moisture blown in from elsewhere. The plants' roots normally absorb water that would otherwise soak through the ground to streams and flow to the sea. Some of that water then returns to the air by evaporating through their leaves. As the region begins to suffer drier summers, however, the plants wilt and possibly die, thereby putting less water back into the air. The vital threshold is crossed when the plant population shrinks to the point that the recycled rainfall becomes too meager to sustain the population. At that point more plants die, and the rainfall diminishes further—in a vicious cycle like the one that turned the Sahara into a desert 5,000 years ago. The region has shown no signs of greening ever since.

Scientists fear they have not yet identified many of the thresholds that, when crossed, would lead to changes in regional climates. That knowledge gap is worrisome, because humans could well be doing many things to tip the climate balance in ways we will regret. Dancing in a canoe is not usually recommended, yet dance we do: We are replacing forests with croplands, which increases how much sunlight the land reflects; we are pumping water out of the ground, which changes how much water rivers carry to the oceans; and we are altering the quantities of trace gases and particulates in the atmosphere, which modifies the characteristics of clouds, rainfall and more.

Scientists fear they have not yet identified many of the thresholds that, when crossed, would lead to changes in regional climates.

Facing the Future

Negative consequences of a major climate shift can be mitigated if the change occurs gradually or is expected. Farmers anticipating a drought can drill wells, or learn to grow crops less dependent on water, or simply cut their losses and move elsewhere. But unexpected change can be devastating. A single, surprise drought year may at first bankrupt or starve only the most marginal farmers, but damage worsens as the drought lengthens—especially if no one had time to prepare. Unfortunately, scientists have little ability to predict when abrupt climate change will occur and what form it will take.

Despite the potentially enormous consequences of a sudden climate transformation, the vast majority of climate research and policymaking has addressed gradual shifts—most notably by calling for global reductions of carbon emissions as a way to slow the gradual rise in global temperatures. Although such reductions would probably help limit climate instability, thought should also be given specifically to avoiding abrupt changes. At one extreme, we might decide to ignore the prospect altogether and hope that nothing happens or that we are able to deal with whatever does happen; business-as-usual did sink the *Titanic*, but many other unprepared ships have crossed the North Atlantic unscathed. On the other hand, we might seriously alter our behavior to keep the human effects on climate small enough to make a catastrophic shift less likely. Curbing global warming would be a step in the right direction. Further investigation of climate thresholds and their vulnerabilities to human activities should illuminate other useful actions.

A third strategy would be for societies to shore up their abilities to cope with abrupt climate change before the next surprise is upon us, as suggested by the U.S. National Research Council. The authors of the council's report pointed out that some former societies have bent in response to climate change when others have broken. The Viking settlers in Greenland abandoned their weakening settlement as the onset of the Little Ice Age made their way of life marginal or unsustainable, while their neighbors, the Thule Inuit, survived. Understanding what separates bending from breaking could prove constructive. Plans designed to help ease difficulties if a crisis develops could be made at little or no cost. Communities could plant trees now to help hold soil during the next windy dry spell, for example, and they could agree now on who will have access to which water supplies when that resource becomes less abundant.

For now, it appears likely that humans are rocking the boat, pushing certain aspects of climate closer to the thresholds that could unleash sudden changes. Such events would not trigger a new ice age or otherwise rival the fertile imaginations of the writers of the silver screen, but they could pose daunting challenges for humans and other living things on earth. It is well worth considering how societies might increase their resiliency to the potential consequences of an abrupt shift—or even how to avoid flipping the climate canoe in the first place.

Crossing the Threshold

By Peter Bunyard
The Ecologist, February 2004

Since 1990 we have experienced the warmest 10 years on record. This has left some parts of the world ravaged by drought and famine, and others suffering freak storms such as those that flooded much of lowland Britain in 2000. France, having experienced a devastatingly hot summer in 2003 then found itself enduring torrential winter rains and unprecedented floods. According to Phil Jones, head of the Climatic Research Unit of the University of East Anglia, the three months of June, July and August 2003 were the warmest ever recorded in western and central Europe. The average temperature for those months was nearly 4° centigrade above the long-term norm and breaking records everywhere—including the UK, where temperatures exceeded the 100° Fahrenheit mark for the first time.

Satellite data reveals that the planet has lost about 10 per cent of its snow cover since the 1960s, and that lakes and rivers in the high latitudes of the northern hemisphere remain frozen for two weeks less than they did one century ago. Glaciers in non-polar regions are also retreating, while Arctic sea ice has not only thinned by some 40 per cent since the 1950s, the surface area that it covers during the spring and summer is also down by up to 15 per cent.

The financial cost of natural disasters in 1998 amounted to $65.5 billion, and the World Health Organisation estimates that the spread of diseases induced by global warming may have led to 5 million deaths. Given that all this is down to a mere 0.6° centigrade increase in global temperatures, what will the future hold?

The Doomsday Alternatives

As climatologists are now certain that it is our greenhouse gas emissions that have led to global warming, we urgently need to know what will happen if we fail to curb our emissions or, worse, continue to add to them. The Intergovernmental Panel on Climate Change (IPCC) has come up with a range of predictions for the next 100 years, all contingent on different scenarios of fossil fuel use.

1. If CO_2 Emissions Remain the Same as They Are Today—375 Parts per Million (PPM) of the Atmosphere

This article first appeared in the February 2004 issue of *The Ecologist*, Volume 34, No. 1
www.theecologist.org.

The truth is that the emissions of yesterday will have their impact tomorrow, and, whether we like it or not, we are committed to further warming—even if we were "magically" to level off our greenhouse gas emissions at the level of today: some 375 parts of carbon dioxide per million parts of the atmosphere (375 ppm). According to such a scenario, global temperatures will rise another 1° centigrade on top of what we have already experienced.

Even that "best" scenario will wreak some havoc. Glaciers and sea ice will in all probability vanish, and the number of extreme climate events, such as floods, landslides, heat waves and violent storms, is bound to increase. Agriculture will be affected, as a lack of rain during the growing season and a spate of heat waves have a catastrophic effect on global food supplies. Worst of it all, as conditions get tougher, we are likely to resort to ever increasing uses of energy, so adding to the potential of global warming in the future.

2. If We Curb Emissions so They Only Rise to 550 PPM

If we could stabilise carbon dioxide concentrations in the atmosphere at about double pre-industrial levels (550 ppm, compared to 280 ppm), global temperatures would rise 2° centigrade over the next 100 years, according to the IPCC. With luck, our current climate system could still cope with such a temperature increase without jumping unexpectedly to a very different and hard-to-predict state.

Nevertheless, we would definitely be committed to substantial sea rises, perhaps a foot or more, as sea water expanded in volume as it got hotter; this would be exacerbated by the further melting of glaciers and polar ice. Increased rainfall, particularly over Siberia, would also lead to a significant increase in the flow of cold fresh water into the Arctic Circle, which would curb the flow of the Gulf Stream and its vital transport of heat from the tropics to the high northern latitudes. We would be subjected to ever stronger climate events, including storms and sea surges, torrential rains and their deadly counterpart—drought.

3. If Energy Use Continues to Grow at the Current Rate

Our insatiable and growing appetite for fossil fuels means we are heading for a fourfold increase in greenhouse gases compared to pre-industrial times. That being so, the UK Met Office's Hadley Centre for Climate Prediction and Research envisages a catastrophic 8° centigrade rise on today's global average. We would be then in a range of global temperatures not seen since 40 million years ago, when the planet had no permanent polar ice sheets and sea levels were 12 metres higher than today.

We would lose our major capital cities and much of our best farmland, and be subjected to violent weather on account of the much greater energy trapped at the earth's surface. The circulation of air and the movement of oceans would be fundamentally different from what we experience today. Survival under such circumstances would most likely be impossible, especially in those parts of the world where we have already ravaged the environment.

4. If We Take into Account Neglected Variables

In its business-as-usual scenario, in which global emissions of greenhouse gases continue to rise uncurbed, the IPCC anticipates that by 2100 the concentration of carbon dioxide in the atmosphere will rise to 700 ppm—double that of today. However, the IPCC's predictions neglect the impact of global warming on soils and vegetation.

Until now most climate models, especially those used by the IPCC, have assumed that carbon dioxide will be drawn down out of the atmosphere at a constant rate; this would offset up to half our current emissions. Such models are inherently deficient and far removed from the real world in which the interchange of gases

---

Scientists who maintain that increased growth of forests in the northern parts of Siberia and Canada will counteract global warming are found to be fundamentally wrong.

---

between the earth's surface and the atmosphere is contingent on living processes such as photosynthesis and respiration.

When there is more photosynthesis than respiration the earth's plant life and soil organisms become a sink for carbon. Such is the situation today. But if respiration exceeds photosynthesis the situation reverses and that store of carbon begins to be consumed; soils and vegetation emit greenhouse gases, and become a source of carbon.

The fact is that the Hadley Centre's climatologists are now finding that the IPCC's climate models (used to inform governments) are far too optimistic in their conclusions. Once different vegetation types (ie, broadleaf trees, tropical forests, savannah and grasslands) are integrated into the dynamic of climate change, there is a very different climate story from that when life is left out of the equation. For instance, those scientists who maintain that increased growth of forests in the northern parts of Siberia and Canada will counteract global warming are found to be fundamentally wrong. Why? Because the boreal forests are quick to shed winter snow on account of their shape, thereby exposing their dark needles to the rays of the spring sun. In contrast to the snow-covered tundra and swamps, boreal forests warm themselves and their surrounding environment when all around is cold. As Hadley Centre climatologist Richard Betts has found, the warming from the dark sun-exposed leaves more than counteracts the cooling that accrues from the growing forest taking up carbon dioxide. Forest growth in the Arctic Circle gives us a warmer, not a cooler, planet.

Currently, one half of all global emissions of greenhouse gases are absorbed into soils and oceans during the course of each year. The growth of forests and storage of carbon compounds in soils therefore play an important role in acting as a "sink" for carbons, thus reducing the overall impact of our emissions. But how permanent is the "storage" of that carbon? Could it suddenly be released back into the atmosphere and become an additional "source" of greenhouse gases, just when the heat is on and we least want it?

Most climatologists base their predictions of future climate change on the grounds that the stores of carbon in soils and vegetation will remain intact as if for ever, and that the sinks for our carbon dioxide emissions will continue to operate come what may. Yet when the Hadley Centre climatologists included carbon cycle feedbacks in their climate models they found that disturbing changes would be likely to occur across the planet.

2080—The Nightmare Scenario

By 2080 the pattern of rainfall would be fundamentally different, with somewhat greater precipitation over the high latitudes—including the ocean. But across the tropics (except for a region in the Pacific) rainfall would decline by 50 per cent or more over all continents. With far less broadleaf forests in the tropics as a result of declining rainfall, daytime temperatures would be likely to rise by a substantial 10° centigrade. That, and the lack of rain, would be devastating for agriculture right across the planet. It would also be devastating for settlements, cities and industry. The corn-belt of the US would suffer from a 30 per cent decline in rainfall during its critical growing season, quite aside from an increase in heat waves. And, with more energy retained within the tropics (especially in the oceans), our coastlines would be battered by violent sea storms, including hurricanes and typhoons, as well as sea surges on a scale we have never seen before: With raised sea levels, the damage inflicted by such storms on vulnerable coastlines, such as along the Ganges Delta or in Indonesia and in Europe, would be unimaginable. Such climatic horrors would trigger a flood of refugees that would make today's numbers appear a trickle.

Switching Off the Gulf Stream

One probable consequence of wetter, warmer conditions in the northern hemisphere is that the Gulf Stream would judder to a halt, or at least shift much further south—taking its warmth with it. Just imagine the consequences of Labrador-like winters over northern France, all of the UK and Scandinavia: an ice sheet would develop, spreading over northern Europe and certainly covering much of Britain, as happened in the last Ice Age of 100,000 years' ago.

> ## *The methane store is a bombshell waiting to go off.*

The Gulf Stream works the way it does because of the saltiness and low temperature of the surface waters in the higher latitudes. The cold, salty water becomes denser than the waters beneath and sinks to the bottom. From there it flows back to the equator and south towards Antarctica. That conveyor belt circulation picks up nutrients on its long passage at the bottom of the Atlantic, and when those same waters, some thousand years' hence, rise back to the surface to become the Gulf Stream, they are rich in essential elements for the growth of plankton. That's why the northern Atlantic provides one of the richest fishing grounds in the world.

But global warming is causing glaciers to melt in Greenland and Canada; it is also causing a substantial increase in rainfall over Siberia. Consequently, the flow of fresh water into the Arctic Circle is diluting the saltiness of the northern waters of the Gulf Stream. At some critical level the surface waters will neither be cool nor salty enough to sink, and a log jam of warm water pushing up from behind will cause the system to stall. Climate records gleaned from ice-core samples and from the ocean bottom show that a similar stalling has occurred in the past.

What has surprised geologists and climatologists is the suddenness with which the flowing Gulf Stream can stall and the temperature can change over northern Europe: it can all happen in a matter of years, not centuries or millennia. Marine scientists from Scotland and the US have found a 20 per cent drop in the temperature of the deep-bottom flow of the "overturned" waters from the Gulf Stream close to the Faroe Islands. Again feedbacks are involved. Less Arctic sea ice means that less light is reflected away during the spring, summer and autumn and more is absorbed into an ice-free sea. That will prevent the Gulf Stream waters cooling sufficiently, let alone retaining sufficient saltiness for sinking to occur.

No one knows precisely the critical turning point at which the system will flip. Could we be on the very edge of it now?

The Methane Time Bomb

Currently, several hundred million tonnes of methane leak into the atmosphere every year; most of which comes from poorly maintained gas pipelines, rice paddies, cattle farming, the draining of wetlands and the destruction of forests. Over the past 250 years, largely because of human activities, methane concentration in the atmosphere has more than doubled to 1.72 ppm. It is now accumulating in the atmosphere at the rate of around an extra 1 per cent per year. Weight per weight, this potent greenhouse gas is 20 times more powerful over a 100-year time span than carbon dioxide.

Fortunately for our climate, most of the methane produced remains trapped a few hundred metres down in the sea as methane hydrate—an ice-like water-methane compound. Methane's majority ingredient is carbon, and the total methane store could constitute as much as 10,000 billion tonnes of carbon—more than 10 times the carbon now found in the atmosphere. The release of just one 10th of that methane store would not only double atmospheric carbon; its impact on global warming would be more than 10 times greater than an equivalent quantity of carbon dioxide.

The methane store is a bombshell waiting to go off. Methane levels in the atmosphere have not been so high since 160,000 years ago, when the earth was undergoing rapid global warming. Could global warming, combined with sea-level rise, suddenly trigger the release of enough methane to raise temperatures far higher than those projected by the IPCC? Most disturbingly, once global warming gets underway more and more methane will vent into the atmosphere. Global warming will beget more global warming.

Dire Warming Warning for Earth's Species

BY JANE KAY
THE SAN FRANCISCO CHRONICLE, JANUARY 8, 2004

More than 1 million plant and animal species will vanish if global temperatures continue to rise as predicted in the next 50 years, scientists say in the first authoritative attempt to gauge the impact of climate change on wildlife.

Even small fluctuations in climate can affect a species' ability to remain in its original habitat and survive, according to a study published Wednesday in the British journal *Nature*.

Authors estimate that about a quarter of the estimated 5 million or more land species on the planet may lose habitat and face extinction as they seek cooler temperatures to survive, either by moving to higher ground or away from the equator and closer to the poles.

The key problem, said co-author Lee Hannah, a biologist at the Center for Applied Biodiversity Science at Conservation International in Washington, D.C., is that vulnerable species have no escape routes as temperatures rise.

"Slight increases in temperature can force (a species) to move toward its preferred, usually cooler, climate range. If development and habitat destruction have already altered those habitats, the species often have no safe haven," he said.

The researchers examined impacts on flora and fauna of three scenarios of climate change between now and 2050, ranging from an average annual temperature increase of about 1.4 to 3.6 degrees Fahrenheit.

The midrange prediction shows temperatures increasing by about 3 degrees. Under that scenario, said lead author Chris Thomas, a biologist at the University of Leeds (England), about 1.25 million species, or 24 percent of the terrestrial species of plants and animals, will be extinct or on the way to extinction in 50 years.

Results were drawn from studies of 1,103 animal and plant species in Mexico, Australia, Europe, the Amazon, the Brazilian Cerrado and South Africa, representing 20 percent of the Earth's landmass.

Researchers estimated that 15 to 37 percent of those species could go extinct or be on the path toward extinction by 2050, depending on which of the three climate assumptions is used. The researchers extrapolated the results to make global estimates.

Computer models used to complete the analysis took into account the ability of different species to move under pressure and the current state of their habitat. Some findings are as follows:

- Using the most conservative estimate of global temperature increases, 18 percent of species would be lost or "committed to eventual extinction." In the dry forest of the Brazilian Cerrado, 66 percent of the plants would be facing extinction.

- Under the midrange prediction, 24 percent of species would be lost. Nearly 70 percent of mammals in South Africa that cannot readily move to alternative sites would vanish.

- With a rise of more than 3.6 degrees, 35 percent of species would be lost. In Queensland, Australia, 85 percent of birds would face extinction. In the Amazon, 87 percent of the plants would vanish.

Researchers from 14 institutions around the world cooperated on the study. It began a year ago, when scientists at Leeds, the Royal Society for the Protection of Birds in Bedfordshire, England, and the National Institute of Public Health and Environment in the Netherlands, among others, realized that separate modeling efforts were under way in six regions to discover how global warming might affect species.

In Queensland, the Boyd's forest dragon would have no choice but to move upslope to find its preferred climate. But because the lizard already is found only near mountaintops, it would have nowhere to go.

In South Africa, the toffeeapple conebush, which bears a fruit that looks like a candy apple, is already diminishing from drought and warmer temperatures.

In Europe, the Scottish crossbill, a rare bird, would have to move to Iceland to find a favorable climate, while the red kite, already heavily threatened by hunting and habitat loss, would lose even more territory.

The Oaxacan swallowtail, one of Mexico's rarest butterflies, which was just discovered in 1975 in the Juarez Mountains, could no longer use the natural habitat protected for it under law.

Despite the uncertainties of the modeling, the study authors said, "We believe that the consistent overall conclusions across analyses establish that anthropogenic (human caused) climate warming at least ranks alongside other recognized threats to global biodiversity."

In many regions, global warming will be the greatest threat, the study said.

The authors recommended an immediate reduction of carbon dioxide and other greenhouse gases, which are believed to be the prime cause of global warming. They also advise that conservation measures must be taken to help prepare for the movement by species to survive.

The new study is the latest of several recent reports on global warming, nearly all suggesting a growing consensus that humans are affecting the climate in ways likely to have some devastating impacts on the environment, such as more extreme weather events, earlier spring thaws and flooding along coastlines. Last month, two top U.S. climate researchers released a report that brings near-unanimous agreement that the warming is caused by combustion of fossil fuels and other industrial emissions. They estimate a 90 percent chance that the world's climate will heat up between 3.1 and 8.9 degrees Fahrenheit by the end of this century.

Predicted are more heat waves and droughts in some regions and heavy precipitation in others. Wildfires, vegetation changes and continuing melting of glaciers, causing flooding and island inundation are forecast.

President Bush has decided not to sign the Kyoto treaty, an international agreement setting deadlines for reducing greenhouse gas emissions. The administration's chief meteorologist, James Mahoney, assistant secretary of commerce, said last month that he believed that natural variation played as much a part in warming as human-caused activities. He cautioned against dire scenarios based on many differing atmospheric models.

Northeast Seen Getting Balmier

By Michael Powell
The Washington Post, December 17, 2001

New England's maple trees stop producing sap. The Long Island and Cape Cod beaches shrink and shift, and disappear in places. Cases of heatstroke triple.

And every 10 years or so, a winter storm floods portions of Lower Manhattan, Jersey City, and Coney Island with seawater.

The Northeast of recent historical memory could disappear this century, replaced by a hotter and more flood-prone region where New York could have the climate of Miami and Boston could become as sticky as Atlanta, according to the first comprehensive federal studies of the possible effects of global warming on the Northeast.

"In the most optimistic projection, we still end up with a six- to nine-degree increase in temperature," said George Hurtt, a University of New Hampshire scientist and co-author of the study on the New England region. "That's the greatest increase in temperature at any time since the last Ice Age."

Commissioned by Congress, the separate reports on New England and the New York region explore how global warming could affect the coastline, economy and public health of the Northeast. The language is often technical, the projections reliant on middle-of-the-road and sometimes contradictory predictive models.

But the predictions are arresting.

New England, where the regional character was forged by cold and long, dark winters, could face a balmy future that within 30 to 40 years could result in increased crop production but also destroy prominent native tree species.

"The brilliant reds, oranges and yellows of the maples, birches and beeches may be replaced by the browns and dull greens of oaks," the New England report concludes. Within 20 years, it says, "the changes in climate could potentially extirpate the sugar maple industry in New England."

The reports' origins date to 1990, when Congress passed the Global Change Research Act. Seven years later, the Environmental Protection Agency appointed 16 regional panels to examine global warming, and how the nation might adapt. These Northeast reports, completed about two months ago, are among the last to be released. (The mid-Atlantic report, which includes Washington, was completed a year ago.)

The scientists on the panels employed conventional assumptions, such as an annual 1 percent increase in greenhouse gases in the atmosphere. They conclude that global warming is already occurring, noting that, on average, the Northeast became two degrees warmer in the past century. And they say that the temperature rise in the 21st century "will be significantly larger than in the 20th century." One widely used climate model cited in the report predicted a six-degree increase, the other 10 degrees.

Within 70 years, New York will have as many 90-degree days a year as Miami does now.

The Environmental Protection Agency summarizes the findings on its Web site.

"Changing regional climate could alter forests, crop yields, and water supplies," the EPA states. "It could also threaten human health, and harm birds, fish, and many types of ecosystems."

Yale economist Robert O. Mendelsohn is more skeptical. He agrees that mild global warming seems likely to continue—but argues that a slightly hotter climate will make the U.S. economy in general, and the Northeast in particular, more rather than less productive. A greater risk comes from spending billions of dollars to slow emissions of greenhouse gases.

"Even in the extreme scenarios, the northern United States benefits from global warming," said Mendelsohn, editor of the forthcoming "Global Warming and the American Economy." "To have New England lead the battle against global warming would be deeply ironic, because it will be beneficial to our climate and economy."

The scientists on the Northeastern panels estimated that Americans have a grace period of a decade or two, during which the nation can adapt before global warming accelerates.

"We will face an increasingly hazardous local environment in this century," said William Solecki, a professor of geography at Montclair State University in New Jersey and a co-author of the climate change report covering the New York metropolitan region. "We're in transition right now to something entirely new and uncertain."

New York City, the nation's densest urban center, is armored with heat-retaining concrete and stone, and so its median temperature hovers five to six degrees above the regional norm. The city, the New York report predicts, will grow warmer still. Within 70 years, New York will have as many 90-degree days a year as Miami does now.

If temperatures and ozone levels rise, the report says, the poor, the elderly and the young—especially those in crowded, poorly ventilated buildings—could suffer more heatstroke and asthma.

But such problems might have relatively inexpensive solutions, from subsidizing the purchase of air conditioners to planting trees and painting roofs light colors to reflect back heat.

"The experience of southern cities is that you can cut deaths and adapt rather easily," said Patrick Kinney of the Mailman School of Public Health at Columbia University, who authored a section of the report.

Rising ocean waters present a more complicated threat. The seas around New York have risen 15 to 18 inches in the past century, and scientists forecast that by 2050, waters could rise an additional 10 to 20 inches.

> *Sea-level rise could reshape the entire Northeast coastline.*

By 2080, storms with 25-foot surges could hit New York every three or four years, inundating the Hudson River tunnels and flooding the edges of the financial district, causing billions of dollars in damage.

"This clearly is untenable," said Klaus Jacob, a senior research scientist with Columbia University's Lamont-Doherty Earth Observatory, who worked on the New York report and is an expert on disaster and urban infrastructure. "A world-class city cannot afford to be exposed to such a threat so often."

Jacob recommends constructing dikes and reinforced seawalls in Lower Manhattan, and new construction standards for the lower floors of offices.

Sea-level rise could reshape the entire Northeast coastline, turning the summer retreats of the Hamptons and Cape Cod into landscapes defined by dikes and houses on stilts. Should this come to pass, government would have to decide whether to allow nature to have its way, or to spend vast sums of money to replenish beaches and dunes. Complicating the issue is the fact that some wealthy coastal communities exclude nonresident taxpayers from their beaches.

"Multimillionaires already are armoring their property with sandbags, but they can't do it on their own," said Vivian Gornitz of Columbia's Center for Climate Systems Research, author of the report's section on sea rise. "You would be asking taxpayers to pay for restoring beaches they can never walk on, and they might demand access."

Farther north, global warming could change flora and fauna, and perhaps the culture itself.

Compared with a century ago, the report notes, ice melts a week earlier on northern lakes. Ticks carrying Lyme disease range north of what scientists once assumed was their natural habitat. Moist, warm winters have led to large populations of mosquitoes, with an accompanying risk of encephalitis and even malaria.

"The present warming trend has led to another growing health problem," the report states, "in the incidence of red tides, fish kills and bacterial contamination."

Hot, dry summer months, the report continues, "are ideal for converting automobile exhaust . . . into ozone." Because winds flow west to east, New England already serves as something of a tailpipe for the nation. The report notes that a study of ozone pollution and lung capacity found that hikers on Mount Washington, New Hampshire's highest peak, ended their treks in worse condition than when they started.

These findings are not definitive. Rising temperatures could exacerbate the effects of harmful ozone—but anti-pollution laws are also cutting emissions.

"There is a little tendency to be alarmist in global warming studies," Kinney said. "We could keep ozone in check."

A warmer New England could help some economic sectors. As oak and hickory replace maples and birch, so commercial forestry might grow. Shorter winters could translate into longer growing seasons, lower fuel bills and less money spent on frost-heaved roads. The foliage and ski industries would suffer, but lingering autumns could bring more tourists and dollars to the coastal towns of Maine and Massachusetts.

"People complain that we'll lose the sugar maple, but 100 years ago, New England was 80 percent farmland," said Yale economist Mendelsohn. "In fact, an entire landscape has shifted in the past 100 years, and most people have no idea it was once so different."

Perhaps—though cold has defined New England for almost 400 years, and some historians caution that the cultural shift could prove disorienting. The region reflects its climate; the literature is austere, the houses stout. For the 19th century naturalists of the region, a clammy southern heat represented moral slackness.

"Surviving winter has become our self-selecting filter," said Vermont archivist Gregory Sanford. "What will we brag about if we live in a temperate zone and go around in Hawaiian shirts and sandals?"

Risk to State Dire in Climate Study

By Miguel Bustillo
Los Angeles Times, August 17, 2004

Global warming could raise average temperatures as much as 10 degrees in California by the end of this century—sharply curtailing water supplies, causing a rise in heat-related deaths and reducing crop yields—if the world does not dramatically cut its dependence on fossil fuels, according to a study by 19 scientists published Monday.

The study, in the *Proceedings of the National Academy of Sciences*, contemplated the consequences of two distinct paths the industrialized world could take in response to a changing climate: maintaining its current reliance on coal, oil and gas, or massively investing in new technologies and alternative energy sources. Burning fossil fuels adds carbon dioxide to the atmosphere, which increases global temperatures by trapping more of the sun's heat.

Using two new computer models on climate change, the study focused exclusively on impacts in California, citing the state's economic importance, diverse climate and longtime reputation as a leader in environmental protection.

The scientists' findings were stark. Human activities already have caused an increase in the amount of gases that contribute to global warming, and as population grows, some further increases are inevitable, the researchers said. Because of that, the state will have to endure not only higher temperatures but significantly longer summer heat waves no matter which path is taken, they warned.

Meanwhile, the Sierra Nevada will receive substantially less snowfall. Much of the state's water comes from mountain snow, and that snowpack could be reduced by 89% if greenhouse gases are not reduced, the study predicted. Rising temperatures could also produce more heavy precipitation in the spring, forcing managers of rapidly filling reservoirs to release water they would prefer to save for dry summer months.

"The state is not set up to deal with what could be a thorny problem over how to deal with shortages and diversion," said Michael Hanemann, director of the California Climate Change Center at UC Berkeley.

Nonetheless, the study concluded that aggressive measures to reduce greenhouse gas emissions could make a dent in the global warming problem.

"The question is, are you going to wait 25 years to solve this, or are you going to act on the vast preponderance of evidence that we are accumulating?" said one of the study's authors, Steve Schneider, co-director of Stanford University's Center for Environmental Science and Policy.

If the world continues to release high levels of heat-trapping gases, California's average statewide temperature is likely to rise 7 to 10 degrees degrees Fahrenheit by the end of the century, the study concluded.

On the other hand, if nations undertake large-scale reductions— which the scientists conceded would require major economic and behavioral changes—temperatures are still likely to rise 4 to 6 degrees by 2100, the study found.

"The choices that we make today and in the near future will determine the outcome of this giant experiment we are undertaking with our planet," said Katharine Hayhoe, an Indiana-based climate consultant who was the lead author of the report. An increase of 7 to 10 degrees "is enough to make many coastal cities feel like inland cities do today, and enough to make inland cities feel like Death Valley," Hayhoe said.

If fossil fuel use is not reduced, the study warned, heat waves in Los Angeles would become six to eight times more frequent, and heat-related deaths would increase five to seven times.

The statewide average temperature, taking in day and night throughout the year, is about 60 degrees. It has slowly risen over the last two decades, climate records show. If it continues rising, scientists say it will exceed the range of historical variation within the next 10 years.

The report was produced by scientists who have specialized in the study of climate change. They include researchers from Stanford, UC Berkeley and the Scripps Institute of Oceanography in La Jolla, as well as government experts from the U.S. Department of Agriculture's Corvallis Forestry Sciences Laboratory in Oregon.

While the findings were largely in accord with previous predictions about global warming in California, some conclusions were more extreme, a fact that some participants attributed to new, more detailed climate modeling.

"They are very dramatic, but we have seen similar numbers before in other studies," said Peter H. Gleick, president of the Oakland-based Pacific Institute for Studies in Development, Environment and Security and a 2003 MacArthur fellow who has been studying climate change since the 1980s.

"I guess the surprise is that even the so-called good news doesn't look so good. Those scenarios look very ugly for California. Every scenario shows California's snowpack going away."

Rising temperatures could also affect the state's multibillion-dollar farming industry, the scientists noted. A particular concern is the Napa and Sonoma wine grape harvest, which experts said could be hurt by even a slight uptick in temperature.

"Under higher temperatures, grapes fall off the vine more quickly," and the quality of the valuable fruit can be harmed, said Chris Field, director of the department of global ecology at the Carnegie Institution. Any sizable increase in temperatures "threatens California's status as the leading producer of wine grapes," he said.

Continental Divide

Why Africa's Climate Change Burden Is Greater

BY SCOTT FIELDS
ENVIRONMENTAL HEALTH PERSPECTIVES, AUGUST 2005

Africa can easily be said to contribute the least of any continent to global warming. Each year Africa produces an average of just over 1 metric ton of the greenhouse gas carbon dioxide per person, according to the U.S. Department of Energy's *International Energy Annual 2002*. The most industrialized African countries, such as South Africa, generate 8.44 metric tons per person, and the least developed countries, such as Mali, generate less than a tenth of a metric ton per person. By comparison, each American generates almost 16 metric tons per year. That adds up to the United States alone generating 5.7 billion metric tons of carbon dioxide per year (about 23% of the world total, making it the leading producer), while Africa as a whole contributes only 918.49 million metric tons (less than 4%). It is cruel irony that, in many experts' opinion, the people living on the continent that has contributed the least to global warming are in line to be the hardest hit by the resulting climate changes.

"The critical challenge in terms of climate change in Africa is the way that multiple stressors—such as the spread of HIV/AIDS, the effects of economic globalization, the privatization of resources, and conflict—converge with climate change," says Siri Eriksen, a senior research fellow in sociology and human geography at the University of Oslo. "It is where several stressors reinforce each other that societies become vulnerable, and impacts of climate change can be particularly severe." She cites the example of the 2002 drought-triggered famine in southern Africa, which affected millions due partly to populations' coping capacity being weakened by HIV/AIDS.

"Climate change could undo even the little progress most African countries have achieved so far in terms of development," says Anthony Nyong, a professor of environmental science at the University of Jos in Nigeria. With climate change has come an increase in health problems such as malaria, meningitis, and dengue fever, he says. This means that the few resources these poor countries have that would have been channeled into essential projects to further economic development must instead be put toward health crisis after health crisis, providing emergency care for the people.

Models of Change

Africa may already be feeling the effects of global warming, says Isaac Held, a senior research scientist at the National Oceanic and Atmospheric Administration (NOAA) Geophysical Fluid Dynamics Laboratory, and more marked effects are likely to come, according to models developed by climate scientists around the world. Projected effects of global warming on precipitation throughout the world can be summarized in a single sentence, according to Held: areas that already get a lot of rainfall—such as the equatorial and subpolar rain belts—will get more, and areas that get little—such as the subtropical dry zones—will get less.

Climate models suggest that subtropical Africa south of the equator will follow this trend, and a plausible case can be made that global warming has already reduced rainfall in that region. In a paper published in the September 2004 issue of the *International Journal of Climatology*, NOAA scientist Pingping Xie and colleagues wrote, "Large decreasing rainfall trends were widespread in the Sahel from the late 1950s to the late 1980s; thereafter, Sahel rainfall has recovered somewhat through 2003, even though the drought conditions have not ended in the region." The study also found that major multiyear oscillations have appeared to occur more frequently and to be more extreme since the late 1980s.

About 300,000 people died in a prolonged drought in the Sahel during the 1970s. Until recently the scientific community attributed that drought to the severe loss of vegetation accompanying such factors as overgrazing and overpopulation; according to this model, the reduction in vegetation meant greater reflectivity of the Earth's surface and less moisture being returned to the atmosphere, with a net drying effect. But now, Held says, "we think of that drought as having been driven by changes in the ocean temperatures."

New climate models posit that precipitation changes are occurring because of alterations in the temperature gradient between the Southern and Northern hemispheres. "When the [ocean] waters are warmer in the Northern Hemisphere, rains are attracted farther north, and when they are warmer in the Southern Hemisphere, it doesn't get as far north," Held explains.

But it is a subject of debate, he says, whether the changes in temperature gradient that caused the Sahel drought were due to natural variability of the oceans or were partly the result of man-made changes in the composition of the atmosphere. There is as yet no consensus among climate modelers on the most likely future trend of Sahel rainfall, he emphasizes.

Still, climatologists agree that warming is happening. In *Climate Change 2001: Impacts, Adaptation and Vulnerability*, the Intergovernmental Panel on Climate Change Working Group II reported, "The historical climate record for Africa shows warming of approximately 0.7°C [1.3°F] over most of the continent during the twentieth century; a decrease in rainfall over large portions of the Sahel . . . and an increase in rainfall in east central Africa."

Accompanying the Sahel droughts has been an increase in dust storms, although whether these storms are a result of the droughts or a cause has been a subject of controversy. Research published in the 22 May 2001 issue of the *Proceedings of the National Academy of Sciences*, for example, presents an analysis that demonstrated that the particles in Sahel dust storms weren't able to absorb much water. As a result, they did not form cloud condensation nuclei, as is usually expected in such dust particles, and so they suppressed rainfall and exacerbated the droughts. In addition to effects on weather, dust motes averaging less than 2.5 micrometers in diameter are in the range of particles that research shows may have serious health consequences. Such dust can carry with it a variety of hitchhikers including bacteria, fungi, and chemical pollutants, all of which may adversely affect health.

Climatologists agree that warming is happening.

Food Stock Failure

On any continent crop failure means trouble, but in Africa it's a catastrophe. About 40% of the gross national product of African countries flows from agriculture, and about 70% of African workers are employed in agriculture, most of them on small plots of land. "Africa is full of poor people who are very highly dependent on climate-related issues for their livelihoods," says Bob Scholes, an ecologist at the Council for Scientific and Industrial Research in Pretoria, South Africa. "They are subsistence farmers in often very marginal environments." African governments are frequently chaotic, ineffective, unstable, and corrupt, adding to the people's precarious existence.

Land ownership changes, less restrictive trade policies, commercialization of the agricultural sector, and increasing impoverishment, along with population growth, have pushed people into farming in dry areas, such as savanna, that not long ago were open to cattle and wildlife grazing, says Jennifer Olson, regional coordinator for land use with the LUCID (Land Use, Change, Impacts, and Dynamics) project at the International Livestock Research Institute in Kenya and a geography professor at Michigan State University. Faced with shrinking open grasslands, once solely pastoral people are settling down and planting crops of their own to supplement their livestock. New farmers tend to be poor, Olson says, and their farms, set in these dry areas, are usually small and thus especially vulnerable to droughts, floods, and other weather hazards associated with climate change.

"Rainfall is the biggest variable for crop and animal production here," Olson says. "Everything goes up and down depending on how the rainy seasons are going, so climate change is going to have a huge impact with the expansion of the number of people doing cropping in the more marginal areas. These tend to be the people on the edge of doing well anyway because there's not enough rainfall for them to be productive."

According to Robert Mendelsohn, a professor at the Yale School of Forestry and Environmental Studies, small farmers in Africa are especially vulnerable to changes in precipitation. Only a small number use irrigation or fertilizer of any kind. Larger growers, such as the commercial farms of parts of East Africa, are better able to cope with weather extremes, but they are in the minority.

People who depend on livestock will be just as hard-hit as pastures go brown, Mendelsohn says. But in this case smaller spreads fare better than large ones. Big operations are usually committed to herds of cattle, which demand plentiful water and easy-to-reach areas in which to graze. When water is scarce, large pastoralists are forced to move their herds southward to relatively wetter areas that are usually occupied by sedentary farmers, thus precipitating intergroup conflicts. However, people who have just a few animals can switch to goats and sheep, animals that tend to be more inventive when it comes to finding food and water.

Ecosystem Changes

With changes in land use and climate, some areas in East Africa have become drier, Olson says, and water sources are becoming intermittent or disappearing. Streams that used to run year-round are now seasonal. The expansion of agriculture into savannas also blocks migration routes for large animals such as zebras, wildebeest, and elephants, she says.

As a result of climate-related ecosystem changes, some wild sources of food are also becoming harder to find, says Catherine O'Reilly, an assistant professor of environmental science at Vassar College. The fish stock in the deep Rift Valley lakes of East Africa, for example, are decreasing as average air temperatures rise. These lakes—a chain of fresh and brackish bodies including lakes Malawi, Tanganyika, and Victoria—contain greater biodiversity than any other of the world's freshwater systems, she says. That diversity depends on algae that are supplied when surface waters mix with nutrient-rich deep waters.

"With climate change," O'Reilly explains, "there is less of this mixing, because the [temperature-mediated] density difference between the surface waters and the deep waters has gotten greater, and so it takes more energy to mix deep water up to the surface." Less algae means less food for the entire food web, and the result, she says, is big decreases in fish catches in all of these lakes.

O'Reilly and colleagues reported in the 14 August 2003 issue of *Nature* that climate change had contributed to a 30% decline in Lake Tanganyika fish stocks over the past 80 years. Such declines can be disastrous for the villages in the region, where the average income is less than US $250 per year, and where the people depend on the fish from these lakes for all of their protein.

When this important food source fades, every aspect of the regional environment is affected. As fish yields go down, increased demands are put upon the land as some fishermen switch to arable farming, O'Reilly says. This in turn leads to more intensified farming, and thus more deforestation, increased erosion, and degradation of the shoreline. Degradation of the shoreline destroys in-shore habitat and spawning grounds for many fish species, further impacting the fish population. "You get a positive feedback loop started," she says, "whereby a small decrease in productivity in the lakes can cycle through all these factors and impact [yet] another aspect of the fish life-history cycle."

When food sources dry up, Africans also turn to wild game. This can put pressure on already endangered species and potentially expose diners to the diseases these exotic animals carry. A report published in the 12 November 2004 issue of *Science* showed that declining fish stocks in Ghana—down by at least 50% since 1970—corresponded with a demand for "bushmeat" that led to a 76% decline in the numbers of 41 species of mammals, including buffalo, antelope, jackals, monkeys, and elephants.

Insects—and with them the diseases they harbor—have also been affected by new climatic conditions, says Jonathan Patz, a physician and associate professor of environmental studies and population health sciences at the University of Wisconsin–Madison. As Africa has warmed, he says, vectorborne diseases—those in which a pathogen is carried from one host to another by pests such as mosquitoes—have increased their range. Malaria, for example, has moved into higher African latitudes as highlands have warmed enough for mosquitoes to breed. (However, other experts, such as Marlies Craig, a malaria researcher at the South African Medical Research Council in Durban, believe that factors besides increasing temperatures, such as increased resistance to drugs, are the cause of this vector spread.) Further, as malaria makes its way into higher latitudes, it reaches people who didn't develop malaria immunity as children. The result is an increase in adult mortality.

Although some areas may become more suitable for some diseases, others may become less so, Patz adds—disease vectors don't universally seek warmer temperatures. Rather, each has optimum conditions in which it thrives. In North America, for example, Patz says the warmer, wetter temperatures that foster some mosquito growth tend to worsen conditions for the tick that carries Lyme disease.

Similarly, research by a team from the Institut Français de Recherche Scientifique pour le Développement en Coopération in Senegal, published in the March 1996 issue of the *American Journal of Tropical Medicine and Hygiene*, indicated that tickborne borreliosis extended its range out of the Sahel and into West Africa most likely because of the Sahel drought of the 1970s.

Wages of War

When the apocalyptic horsemen of famine and pestilence appear, war can't be far behind. Decreasing pastoral lands, decreasing available tillable land, decreasing wild game, and decreasing available water all add up to more strife, Scholes says. "Subtropical dry, arid areas are going to be a huge source of conflict over the next half-century because we still have very, very high population growth rates in those areas, very low economic growth rates, and deteriorating environment," he says.

"Basically," he adds, "not only are the spillover effects environmental, in terms of dust storms and soil erosion and so forth, but there is also massive spillover of people moving out of [more

---

When the apocalyptic horsemen of famine and pestilence appear, war can't be far behind.

---

stressed areas] into better resourced areas." In relatively developed countries such as Nigeria and South Africa, 30% or more of the population consists of illegal immigrants.

Farmers, pastoralists, and the new agro-pastoralists are already competing for water and suitable agricultural and grazing land, Olson says; regional warming and drying can only be expected to worsen the situation. On occasion, she says, the conflicts that result from this competition can turn violent, although most are settled peacefully.

But according to Eriksen, extended periods of increased violent raiding in parts of East Africa have led to loss of livestock and land, and have driven people into a state of destitution that makes them extremely vulnerable to drought events. "Although many conflicts are politically instigated and driven by underlying economic inequities in resource access, rather than climate change as such, increasing drought stress can exacerbate conflict and violence," she says.

Strategies for Coping

As a reaction to so much bad news, in July 2005 the G8 countries—Canada, France, Germany, Italy, Japan, Russia, the United Kingdom, and the United States—approved a pledge to forgive about US $40 billion of debt owed by 18 nations, including 14 African countries. These forgiven debts represent about one-sixth of the total owed by African nations to the G8 nations and international lenders. According to Nyong, relief from these debts will allow these African governments to spend more on local issues related to climate change. The G8 leaders also committed to double annual African aid from US $25 billion to US $50 billion.

But debt relief isn't an instant cure. "We really have to spend some time exploring the implications of [debt relief] for rural economies and urban economies," says David Campbell, a geography professor at Michigan State University who has studied East African communities for more than 25 years. It's also important, he says, to determine what investment should be made with financial aid to maintain the resilience of societies, both urban and rural, in the face of potential increased climatic variability.

Industrialized nations bear another responsibility, Nyong says: "Just as Africa is trying to adapt to these adverse impacts of climate change, the developed countries, particularly the G8 countries, should put in place a mechanism to which they are committed to substantially reducing their greenhouse gas emissions. [With Nigeria] having signed the Kyoto protocol, we want to see definite plans articulated to achieving the targets set by the protocol."

Mendelsohn lays out the bottom line: "As the net income of this land deteriorates, it's not going to be able to sustain the number of bodies that are on it anymore. So the question is, where will these bodies go?" One long-term answer is to try to increase industrialization in Africa to give people other alternatives, to move away from climate-sensitive livelihoods and industries toward those that are not climate-sensitive—ecotourism has been suggested as one possible replacement for farm income.

In the meantime, Campbell says, "It's important that the G8 outcome be committed over the long term to maintaining [financial aid] policies." This financial aid, he says, will be vital to the lives of Africa's poor, who represent an ever increasing segment of African society. And climate change is likely to accelerate such societal stratification, he says. People who have at least some wealth will be better able to switch to different crops, buy a different kind of livestock, or combine growing and herding. "Herders who had taken on farming appeared to be less vulnerable to drought than the people who had maintained themselves in terms of subsisting almost entirely on herding," Campbell says. "So that diversification showed itself to be successful in terms of allowing them to cope with prolonged drought." However, those Africans who don't have sufficient wealth to buffer the effects of increasing climatic variability will plunge deeper into poverty.

But as much as financial aid is needed, Nyong says, the reality is that no amount of money is going to stop climate change from affecting Africa in profound and unpredictable ways. Africa wasn't able to prevent the buildup of greenhouse gases, he says, "What we are left to do now is to adapt to the buildup.

IV. What Can Be Done?
Kyoto and Beyond

Editor's Introduction

Over the years, various measures have been proposed to address the global-warming threat. The articles in the final chapter, "What Can Be Done? Kyoto and Beyond," describe a number of these initiatives. Though such proposals offer some degree of hope, many scientists believe that it is already too late to prevent massive climate change. Moreover, with all the political, social, and economic impediments, the prospect of a truly far-reaching and comprehensive global-warming strategy being implemented any time soon seems increasingly remote.

The Kyoto Protocol, the international climate-change pact, is examined in the first article, "Kyoto Update," by Lynn Elsey. Given its broad reach and the difficult circumstances surrounding its ratification, the protocol is a truly unprecedented achievement. Nevertheless, the United States, the world's leader in greenhouse-gas emissions, is not party to Kyoto, which severely limits its effectiveness. Skeptics also suggest that the pact does not go nearly far enough in reducing emissions and is unlikely to have anything more than a marginal impact on the pace of global warming.

As Paul Tolmé shows in the subsequent piece, "It's the Emissions, Stupid," individuals can greatly reduce their personal contribution to climate change without detracting from their quality of life. Indeed, by employing flourescent lightbulbs, fuel-efficient automobiles, solar panels, and other energy-saving measures, consumers can help the environment and save money.

In the following entry, "Oil Project Goes Underground for Cleaner Air," Gary Polakovic looks at an experiment in Canada in which researchers are storing carbon dioxide underground. While the process—carbon sequestration—is not yet cost-effective, many expect it to emerge as a viable means of reducing greenhouse gas emissions in the near future.

One innovative proposal for combating global warming employs a free-market approach. The recently adopted Kyoto Protocol has mandated a carbon-emissions exchange through which energy producers can sell or trade emissions credits, and similar initiatives are expected to take shape in the years ahead. Brad Foss details this emerging strategy in "Carbon Emissions Trading Is New Weapon to Battle Global Warming."

With the growing concern over fossil-fuel consumption, nuclear energy would seem poised to make a comeback, given that it is essentially emissions-free. However, as Richard A. Meserve observes in "Global Warming and Nuclear Power," there remain considerable impediments to a nuclear renaissance, among them public concerns about the safety of the technology and the vulnerability of nuclear power plants to terrorist attack.

In the climate change debate, the hydrogen fuel cell is frequently put forward as a potential cure-all that could render all other proposals moot. However, in "Is Hydrogen the Solution?" the National Resources Defense Council argues that the technology is unlikely to be put into widespread use any time soon, and consequently other strategies must be implemented before the global-warming situation worsens.

While the federal government has yet to take a concrete stand on climate change, a number of states have crafted their own legislation to counteract the problem. Jennifer 8. Lee discusses some of these initiatives in the final entry, "The Warming Is Global but the Legislating, in the U.S., Is All Local."

Kyoto Update

By Lynn Elsey
WEATHERWISE, May/June 2005

The much-debated Kyoto Protocol went into force on February 15, 2005. However, without the participation of the United States and developing countries, the protocol's goal of stemming the tide of global warming by reducing greenhouse gas emissions remains questionable.

The Kyoto agreement sets binding targets on the signatory nations to reduce greenhouse gas emissions by an average of 5.2 percent below 1990 levels during the commitment period of 2008–2012. Critics contend that the reductions are insufficient to have a meaningful impact. And, more importantly, that without the inclusion of developing nations, who are responsible for a growing level of emissions, and the United States—which currently leads the world in greenhouse gas emissions—the agreement will have little, if any, effect on reducing global warming.

Nuts and Bolts

The industrial nations that have signed the agreement have a variety of targets for reducing their greenhouse gas emissions below 1990 levels. For the European Union overall, the target is 8 percent, but the actual figures vary by country. The United Kingdom's reduction target, for example, is 12.5 percent. Germany elected to take on a larger burden and committed to a 21 percent reduction. In the southern hemisphere, New Zealand has a target of reducing just to 1990 levels during the protocol's timeframe.

The protocol includes a system of trading and credits, so that those who exceed their targets can purchase extra credits from those who reduce their emissions by more than their allocated amount. Companies can obtain extra credits by partaking in various environmentally friendly projects, such as helping developing countries build "clean" power plants or becoming involved in reforestation projects.

Outside the Box

The lengthy and sometimes acrimonious discussions leading up the passage of the Kyoto Protocol may have prompted a number of businesses, states, and even nations to voluntarily enact individual

Weatherwise, Vol. 58, No. 3, pp 16–17, May/June 2005. Reprinted with permission of the Helen Dwight Reid Educational Foundation. Published by Heldref Publications, 1319 Eighteenth St., NW, Washington, DC 20036-1802. Copyright © 2005.

greenhouse gas and other harmful emission cuts, whether as a pre-emptive action—assuming that some type of emissions cuts would be inevitable—or from basic financial savvy. For example, DuPont pledged to reduce greenhouse gas emissions by 65 percent of 1990 levels by 2010; by 2002 the company had actually reduced them by 67 percent. In 2002 the Royal Dutch/Shell Group also met their goal of reducing greenhouse gas emissions by 10 percent from their 1990 levels.

As Massachusetts Governor Mitt Romney has noted, "the same policies that protect the climate also promote energy efficiency, smart business practices, and improve the environment in which our citizens live and work."

In 2003, the Commonwealth of Massachusetts enacted its "no regrets" policy toward climate change: rather than becoming bogged down in the debate about the causes and impact of human activity on climate change, the Commonwealth decided to focus on action.

Enactment of the Kyoto Protocol has already created new fissures in the international arena.

The plan involves coordination between the state's energy, environment, housing, and transportation agencies. Goals include reducing greenhouse gas emissions to 1990 levels by 2010, and below 1990 levels by 2020. Other targets include a commitment by the state to spend up to $17 million to purchase renewable energy; to acquire clean, fuel-efficient vehicles for the state fleet; to promote low energy traffic signals and more efficient night lighting; to create a CO_2 registry with other states; and to create an emissions banking and trading program.

"If we learn decades from now that climate change isn't happening, these actions will still help our economy, our quality of life, and the quality of our environment," Governor Romney said.

The Fallout

Enactment of the Kyoto Protocol has already created new fissures in the international arena. Concerns were recently raised in Australia that the country's decision to opt out of Kyoto could have a negative impact on business. Fears have been raised that exporters, for example, could be saddled with extra taxes from countries that are part of the pact. Additionally, Australian financiers have complained about a potential loss of economic benefits, claiming that exclusion from the agreement inhibits Australians from reaping any monetary benefits from actions, such as planting trees, which would otherwise be eligible for carbon credit trading.

The backlash against the United States for its refusal to take part in Kyoto continues. Lord May, the president of Britain's Royal Society and former chief scientific adviser to the British government, recently said that the Royal Society had calculated that the 13 percent rise in greenhouse gas emissions from the United States between 1990 and 2002 will dwarf all of the cuts and benefits generated from the enactment of Kyoto.

"The focus must now be on setting targets for the period beyond the first phase of the Kyoto Protocol, and which include all countries, both in the developed and developing world," May said during a presentation at the British Embassy in Berlin in March.

May also called on the United States and other governments to accept the need for making cuts in greenhouse gas emissions and warned against political leaders acting as "modern day Neros over climate change, fiddling while the world burns."

"It is essential that the G8 summit [scheduled to take place in Scotland in July] focuses on securing from the United States an explicit recognition that the case has now been made for acting urgently to avoid the worst effects of climate change by making substantial cuts in greenhouse gas emissions," May added.

What Next?

Presently, there are no plans for action after Kyoto expires in 2012. And, according to numerous reports, the prognosis for a continuing international cooperative agreement is not good. Following a recent global conference on climate change in Argentina, Nigel Purvis, a scholar at the Brookings Institution and a senior climate change negotiator in the Clinton and Bush administrations, reported that "the Buenos Aires gathering, however, turned into something of a wake as it became apparent that few nations are prepared to extend Kyoto's targets beyond their 2012 expiration date. Like Washington, D.C.'s cicadas, Kyoto took years to hatch, received enormous attention, and will be short-lived."

Purvis and other observers feel that one of Kyoto's biggest legacies will have nothing to do with reducing greenhouse gas emissions, but rather, with altering the relationship between the United States and the parts of the world that are actively supporting Kyoto and climate change action. "Kyoto demonstrates that America's allies are increasingly shaping the international agenda without it," Purvis said in the *International Herald Tribune*. "Kyoto also illustrates the fact that the international community questions more than ever America's moral authority and its commitment to universal values."

It's the Emissions, Stupid

By Paul Tolmé
National Wildlife, April/May 2005

When Larry Kinney moved into a leaky 20-year-old house in Boulder, Colorado, he did more than unpack boxes, lay down rugs and arrange the furniture. He also slashed the home's energy use to minimize his expenses and impact on global warming.

A senior researcher for the Southwest Energy Efficiency Project, Kinney knew just what to do: First, he sealed and reinsulated the house, finding and curing air leaks. Then he opened an internal door and window at the top of a solar-heated space above the main house, so warm air would flow down into the home on sunny winter days. Inside the solar space, he installed a clothesline and a south-facing skylight that adds warmth on winter days and ventilation on hot summer nights. He installed dimmers on light switches that control incandescent bulbs and replaced some 20 incandescents with compact fluorescents. To an existing whole-house fan, he added controls that automatically turn it off when cooling isn't needed and an insulated shutter that seals the fan opening in winter.

Small steps perhaps, but the impact was dramatic. Several years and $3,500 later, Kinney and his wife have a home that costs just $50 a year for heating, cooling and hot water. And because most of their energy comes from the sun, they add little heat-trapping carbon dioxide (CO_2) to the atmosphere.

As global temperatures continue to rise due to humanity's reliance on fossil fuels, Kinney exemplifies how some Americans are becoming part of the solution. By maximizing the efficiency of their homes and using renewable energy, consumers can help decrease the nation's unhealthy dependence on fossil fuels.

Of course, reversing the mounting effects of global warming also requires official action at both national and international levels. In the United States, "we need strong policies such as those included in the proposed Climate Stewardship Act," says NWF climate change program manager Jeremy Symons. Such policies would require power plants and factories to reduce emissions, raise fuel efficiency for automobiles and other motor vehicles, eliminate subsidies for the fossil fuel industry and invest in renewable energy and energy-efficient technologies.

Energy Guzzlers

The country has a lot of work to do. Together, U.S. citizens emit 21 tons of heat-trapping gases per capita—twice the amount of Western Europeans who enjoy the same standard of living—largely because of the energy burned in inefficient homes and gas-guzzling vehicles. Motor vehicles alone are responsible for nearly a quarter of the country's annual CO_2 emissions. Though the United States has just 4 percent of the world's population, it is responsible for a quarter of all global carbon dioxide emissions. And greenhouse gases emitted today will linger in the atmosphere for decades.

Faced with such daunting challenges, people may feel helpless. "It's such a big problem that people say, 'Wow. It's impossible for me to make a change,'" says Stephen Saunders, president of the newly formed Rocky Mountain Climate Organization. "Untrue."

According to the Union of Concerned Scientists, if every U.S. family replaced one regular incandescent bulb with a compact fluorescent, the nation could decrease CO_2 emissions by more than 90 billion pounds annually—equivalent to taking 7.5 million cars off the road. If every U.S. household used the most efficient appliances, the nation could save $15 billion in energy costs and eliminate 175 million tons of greenhouse gases, notes the American Council for an Energy Efficient Economy. "Every time we make a decision to buy a high-efficiency appliance, we're not just reducing our own energy use and saving money, we are sending a powerful signal to manufacturers that people want these products," says Harvey Sachs, director of building programs for the organization.

Though the United States has just 4 percent of the world's population, it is responsible for a quarter of all global carbon dioxide emissions.

Smart Investments

Energy-efficient products, such as those endorsed by the federal government–sponsored Energy Star program, cost more up front but deliver investment returns to rival the best stocks. Compact fluorescent bulbs, for example, cost about $4.50 versus 50 cents for a conventional bulb. But the long life and efficiency of compact fluorescents means they save $5 per bulb every year. Swap 10 bulbs, and the savings are $50 per year. Over the eight-year life of those bulbs, the savings would be $400. "It's obscene not to buy them," says Sam Rashkin, national director of EPA's Energy Star For Homes program, who calculates the rate of return on a compact fluorescent at 110 percent.

Similarly, an Energy Star clothes washer costs about $300 more than an inefficient model, but will save roughly $90 per year in electricity costs, or $800 over the 15-year life of the washer—a 15 percent annual rate of return. "These are investments," Rashkin says. "But the returns don't depend on whether the stock market is good or bad. They are guaranteed, because energy costs are only going to rise."

Demanding renewable energy is another way consumers can influence the market, reducing both their energy expenses and emissions of heat-trapping gases. Half of U.S. households have the option to sign up for green pricing programs, for example, yet only 500,000 consumers do (see "Solutions," below). Still, renewable energy is gaining steam: Between 1995 and 2003, at least $124 billion was invested in installing new renewable energy facilities such as wind farms and solar collectors worldwide, including $24 billion in 2003, says researcher Eric Martinot, a fellow with the World Watch Institute. Within ten years, the renewable energy market is expected to reach $85 billion annually, according to Martinot's report, *Global Renewable Energy Markets and Policies.*

Blowing in the Wind

Wind has the greatest potential. The price of wind power has dropped 40 percent since the 1980s and is now one of the cheapest energies to produce. In the United States, about $2 billion is spent annually ($9 billion worldwide) on building wind farms, which generate enough power to supply 1.6 million homes. Wind constitutes just 1 percent of the electric market in the United States, but the figure could easily grow to 6 percent by 2020, says the American Wind Energy Association (AWEA). Denmark illustrates the technology's promise. In just over a decade, the country went from virtually no wind energy to supplying 20 percent of the country's needs today. "The really good news is that people can no longer hide behind the old argument that wind is expensive and will wreck the economy," says AWEA spokeswoman Christine Real de Azua. "Technologically and economically, converting to wind is feasible."

Solar power, while more expensive to produce than wind, is heating up as well. Global production of energy produced by solar panels in 2004 was about 700 megawatts—worth roughly $3.5 billion and up from 60 megawatts in 1996. In some places, photovoltaic panels are in such demand that there are waiting lists of eight months or more.

Utility Goes Green

Among a handful of utilities leading the charge is Austin Energy, which serves 700,000 customers around the Texas capital. Thanks to its investments in West Texas wind farms, the company sells more wind power than any other in the country. While renewables constitute just 5 percent of the utility's power supply today, it plans to raise that figure to 20 percent by 2020. "The price of fossil fuels is going to continue to rise," says Roger Duncan, the company's deputy general manager, "so wind just makes good business sense."

Two decades ago, Austin Energy had planned to build a new coal-power plant, but Duncan—then a city councilor—helped implement a conservation plan that offset the need for more coal power. In essence, the company "built" a virtual power plant through efficiencies. In 2005, the company plans to spend $20 million on conservation. Looking forward, it would like to fill 15 percent of its energy needs through further efficiency measures. "It's cheaper to save a megawatt of energy through conservation than it is to build new power plants," says Duncan.

Likewise, it would be cheaper for the United States as a whole to reduce its reliance on oil than to continue importing crude from around the globe. According to the Rocky Mountain Institute, eliminating oil imports entirely would pump $70 billion into the national economy. "Saving and substituting for oil costs less than buying it," says the group's CEO, Amory Lovins, who adds that these savings act like "a giant tax cut for the nation. It simply makes sense and makes money for all."

Refuting Skeptics

Indeed, contrary to critics who say that reducing dependence on fossil fuels would wreck the economy, investments in renewables and efficiencies would spur an economic revolution similar to the birth of the Internet and high-tech boom of the past two decades, says economist Michelle Manion of the Union of Concerned Scientists. "We have done a good job refuting the science skeptics" who dismiss the reality of global warming, she says. "Now we need to do a better job refuting the economic skeptics."

Supporting that goal, farsighted individuals like Larry Kinney provide examples of economic benefits that accrue from lowering fossil fuel use. Another Colorado resident, Will Toor, is also doing his part. After moving into a new home in 2004, the Boulder County commissioner replaced incandescent light bulbs with compact fluorescents; sold the old power-hogging appliances and bought an energy-efficient refrigerator, dishwasher and washing machine; replaced the dryer with a clothesline; and signed up for a green-pricing program with his electric company, which supplies 100 percent wind power for an extra $9 a month. The result: His family's monthly electricity use dropped by two-thirds from the previous owners.

Toor, who walks, rides the bus or bikes to work, is proud to be part of the solution. Though he knows he will save money over the long term, he's never bothered to crunch the numbers. Rather, Toor minimized his greenhouse gas emissions because doing so is morally sound. "It's all part of teaching our children to tread lightly on the Earth," he says. "Our children aren't responsible for global warming, but they are the ones who will have to grapple with our mistakes."

Solutions: Building Green

The energy consumed in the average U.S. home produces nearly 10,000 pounds of greenhouse gas emissions per year. Office buildings, which efficiency experts compare to a fleet of SUVs across the nation's skyline, are also energy guzzlers. But thanks to a green building push, new homes and buildings are in for an energy conservation makeover.

Last fall, San Francisco became the ninth city to adopt a Green Building Ordinance that requires all new building construction projects and additions to be certified by the U.S. Green Building Council's Leadership in Energy and Environmental Design program, or LEED. Buildings are rated as Certified, Silver, Gold or Platinum, depending on whether they are "pale" or "dark" green.

The headquarters of the municipal water agency for Chino, California, exemplifies dark-green construction. The building includes rooftop solar panels that provide a quarter of the building's energy. A majority of the construction materials were made locally, reducing the "embedded energy" expended in their manufacture and transport. Three-quarters of the workspaces use natural light. To discourage car commuting, employees have bike lockers, showers and charging stations for electric vehicles, as well as a carpool program and high-mileage alternative fuel vehicles for day trips.

As for homes, an Energy Star program for homes—sponsored by the U.S. Department of Energy and U.S. Environmental Protection Agency (EPA)—already labels as energy efficient both new and existing houses that meet the agencies' specific criteria. See www.energystarhomesamerica.com. In addition, this year a pilot LEED program, to be followed in 2006 by a permanent program, "will be the first step toward ramping up the number of energy efficient homes on a national basis," says Ann Edminster, co-chairwoman of the committee devising the standards. Surveys show that Americans are clamoring for efficient homes, which have a higher resale value. "This is a big market opportunity," Edminster says. "We hope to get to the point where green homes are the rule rather than the exception." See www.usgbc.org.

Solutions: Green Pricing Programs

Half of all U.S. electricity consumers could buy renewable energy from their utilities if they chose to spend a little extra money. But only half a million do. The problem is that most consumers don't even know the option is available. "It's difficult to get the word out," says Lori Bird of the National Renewable Energy Laboratory. "Even the programs that have been around for a while struggle with awareness." More than 500 utilities in 34 states offer green pricing programs. For information, see www.eere.energy.gov/greenpower/markets/pricing or www.green-e.org. You can also support clean energy through NWF's "Green Tag Club." By joining the club, you'll prevent the annual release of tons of the emissions that cause global warming. See www.nwf.org/energy.

Solutions: Turn Your Home into a Power Plant

In the past, only off-the-grid rural residents put up solar panels or wind turbines. But thanks to new technology, state rebate programs and more favorable laws, homeowners can now feed power into the grid and watch their electricity bills shrink.

For wind power options, check out Southwest Windpower's Whisper Link system, which includes a turbine with a nine-foot blade. With a 30- to 50-foot tower, the system costs about $7,000 installed. (Make sure local zoning laws allow the turbines.) In places with average winds of 12 miles per hour such as the Midwest, or in areas with high electricity costs such as Southern California, the system will pay for itself in as little as three years with rebates, according to vice president Andy Kruse. "These are ideal for homeowners with half an acre or more." See

---

State and community officials are implementing Climate Action Plans, pledging to meet the goals of the Kyoto Protocol by cutting emissions of greenhouse gasses.

---

www.windenergy.com. For solar, BP Solar is the world's largest producer of the technology, with more than 30 years of experience. Its blue, dark-framed solar modules are backed by a 25-year warranty. See www.bpsolar.com/homesolutions. Unfortunately, some states do not have "net metering" laws that require utilities to buy back home-generated power for the same price the utilities sell it, which discourages residential wind and solar power generation.

Solutions: States and Communities Take Action

When Oregon set limits on CO_2 emissions from new power plants in 1997, the world paid little notice. But Oregon's bold action began one of the most hopeful trends in climate protection—state and local action. While the federal government drags its feet, state and community officials are implementing Climate Action Plans, pledging to meet the goals of the Kyoto Protocol by cutting emissions of greenhouse gases.

Last September, for example, New York's Public Service Commission adopted a rule that requires 25 percent of the state's electricity to come from renewable resources by 2013. The move is expected to jump-start efforts to generate wind, solar and tidal power, and also capture landfill gas and increase the use of biomass (wood chips and other plant products). Already, 18 states have adopted "Renewable Portfolio Standards" that require utilities to add renewables to their transmission grids. If all 50 states adopted similar standards, renewable energy would boom.

Hoping to turn their blustery and sunny states into the Saudi Arabia of wind and solar power, the Western Governors' Association in 2004 approved a resolution to increase renewable energy production, which would require 30,000 megawatts of clean energy to be produced by 2015 and encourage energy efficiency gains of 20 percent by 2020. Meanwhile, California's decision to force automakers to cut CO_2 emissions from cars and trucks could reshape the auto industry. To meet California's demands, car makers likely will have to improve fuel efficiency across their fleets if they want to sell vehicles in the state—one of the world's largest markets for automobiles.

Frustrated by federal inaction, cities and towns are tackling global warming as well. Seattle, Portland, San Diego, Salt Lake City, Austin and Minneapolis are among large cities that have implemented programs to cut emissions, as have Boulder and Fort Collins, Colorado, Burlington, Vermont, Cambridge, Massachusetts, and New Haven, Connecticut. Chicago and Los Angeles have adopted Climate Protection Programs. San Francisco plans to reduce its greenhouse gas emissions by more than 2.5 million tons by increasing mass transit and hybrid vehicles, implementing energy conservation measures, requiring green building methods and installing solar power collectors on buildings and homes. Seattle's municipally owned electric utility has adopted a climate-neutral program through which it invests in emissions reductions programs around the world to offset its own CO_2 output.

In all, more than 150 communities across the country and 600 worldwide have joined the Cities for Climate Protection program (www.iclei.org/us/ccp). "At first we had to solicit communities," says director Abby Young. "Now we have cities coming to us. Until we get some leadership at the federal level, communities and states must keep the pressure on."

Reduce Your Own Greenhouse Gas Emissions

Reducing your family's heat-trapping emissions does not mean giving up modern conveniences—it just means making smart choices and using energy efficient products. Here are ten easy steps suggested by the Union of Concerned Scientists:

Pick the Right Car—When you buy your next car, look for the one with the best fuel economy in its class or consider new technologies like hybrid engines. Each gallon of gas you use releases 25 pounds of heat-trapping CO_2.

Choose Clean Power—Power plants are the single largest source of heat-trapping gases in the United States, but in some states you can switch to utilities that provide 50 to 100 percent renewable energy. See www.green-e.org.

Look for Energy Star—When it's time to replace appliances like refrigerators, furnaces, air conditioners and water heaters, ask for products with the Energy Star label, which use less electricity than traditional appliances.

Unplug a Freezer—Unplugging just one rarely used refrigerator or freezer can reduce a typical family's CO_2 emissions by nearly 10 percent.

Get a Home Energy Audit—Many utilities offer free audits, which may reveal simple ways to cut emissions—sealing and insulating heating or cooling ducts, for example.

Choose Efficient Lighting—If every U.S. family replaced one regular incandescent bulb with an efficient compact fluorescent, we could decrease emissions by more than 90 billion pounds annually—equivalent to taking 7.5 million cars off the road.

Think Before You Drive—If you own more than one vehicle, use the less fuel efficient one only when you can fill it with passengers. If possible, join a carpool, bike or take mass transit.

Buy Good Wood—Buy certified wood to support sustainably managed forests, which can store carbon more effectively. See www.fsc.org and www.aboutsfi.org.

Plant a Tree—In addition to storing carbon, trees in urban areas and near residences provide summer shade, reducing energy bills, fossil fuel use and greenhouse gas emissions.

Let Policymakers Know You Care About Global Warming—Both elected officials and business leaders need to hear from concerned citizens like you.

Oil Project Goes Underground for Cleaner Air

BY GARY POLAKOVIC
LOS ANGELES TIMES, FEBRUARY 15, 2004

The Saskatchewan prairie is so featureless and flat that people here say you can watch your dog run away for three days. Fields of canola and wheat fill the vastness before falling off the horizon in a landscape punctuated by oil pump jacks bobbing lazily like old men in rocking chairs.

But deep underground, an ambitious experiment is underway to determine whether carbon dioxide can be safely buried. If so, carbon sequestration, as the process is called, could prove to be an effective way to reduce one of the biggest contributors to global warming.

Carbon dioxide emissions from power plants, auto tailpipes, factories and other sources are contributing significantly to global warming, scientists have found. Without sizable reductions in this and other so-called greenhouse gases, the trend is expected to continue, bringing with it tumultuous weather patterns, melting glaciers and rising sea levels.

The Weyburn oil field, 70 miles south of Regina and 50 miles north of the U.S. border, can hold an estimated 21 million tons of carbon dioxide over the project's 25-year lifespan. Saskatchewan's oil fields have enough capacity to store all the province's carbon dioxide emissions for more than three decades, according to the Petroleum Technology Research Center, which manages the project.

The Canadian government believes that carbon gas storage will go a long way toward helping the country meet its emissions reduction targets under the 1997 Kyoto Protocol, of which it is a signatory. The pact requires industrialized nations to cut emissions of greenhouse gases by an average of 5% between 2008 and 2012.

The Weyburn project began four years ago and has the backing of international energy companies and the United States, European Union and Canada, which have contributed $21 million.

"When I first heard about this, my reaction was, 'Wow, this is really nuts.' But the more I looked at it, I began to start believing that, on a small scale, this is something that is achievable," said Sally Benson, deputy director at Lawrence Berkeley National Laboratory in Northern California.

An essential ingredient of the atmosphere, carbon dioxide keeps the Earth warm enough to sustain life. It is only during the last century that scientists have begun to fear that the industrial world is creating too much of it and started to look for ways to lower CO_2 production or dispose of the excess.

> *[Carbon dioxide] is a tricky molecule to get rid of. It doesn't always stay buried.*

But it is a tricky molecule to get rid of. It doesn't always stay buried.

In 1986, 1,700 people in the West African nation of Cameroon suffocated when a giant bubble of naturally occurring carbon dioxide suddenly erupted from Lake Nyos and displaced all of the available oxygen in the immediate area.

Deep-well injection of the gas can force briny water to the surface, potentially polluting streams and aquifers. Earthquakes have been reported in places where deep-well injection has occurred. And carbon dioxide can convert to an acid in groundwater.

"With geological sequestration, we need to know that the carbon we put in the ground isn't going to come back up," said Klaus Lackner, a scientist with the Earth and Environmental Engineering Department at Columbia University.

Burying the gas is one of several remedies for global warming that include energy conservation, emissions reductions and greater reliance on alternative energy. But carbon storage offers a unique incentive. Buried in an oil field, the gas boosts oil production by forcing residual deposits to the surface. At Weyburn, oil production is up 50% since carbon dioxide injection began four years ago.

The Weyburn site was selected because, during 44 years of oil exploration, Saskatchewan required oil companies to keep copious geological records. Core samples from 1,200 bore holes provide a comprehensive look at subsurface conditions and a way to track movement of oil and gases. Carbon dioxide is injected nearly a mile underground beneath a thick rock layer.

Researchers keep track of buried carbon dioxide by checking vapors in wells, sampling groundwater and conducting seismic tests that depict subsurface conditions. So far, no leaks have been detected and none of the gas has escaped to the surface, said Mike Monea, who manages the Weyburn project for the Petroleum Technology Research Center.

But the Weyburn site is pocked with hundreds of oil wells over a 70-square-mile area. Each well shaft can act as a conduit to channel carbon dioxide to the surface. Some wells are being closed off, and others are being watched for traces of carbon dioxide. Scientists are trying to forecast how the site will hold up over several millenniums. One computer model showed that carbon dioxide could migrate upward about 150 feet in 5,000 years—though it would still be far below the surface. A final report is due in June.

Each day, about 5,000 tons of liquefied carbon dioxide arrives from a plant near Beulah, N.D., operated by the Dakota Gasification Co., which converts coal to natural gas. The liquid CO_2 crosses the prairie in a 220-mile-long pipeline before it is pumped underground in Canada.

Separating carbon dioxide from other exhaust exiting a smokestack is expensive. The process can use up to one-third of the energy produced by the power plant. Scrubbers must be installed, pipelines must be built, carbon dioxide must be carefully watched underground.

It costs about $30 a ton to separate carbon dioxide from industrial exhaust, though technology exists to cut that expense nearly in half, said Curt White, carbon sequestration science leader for the U.S. National Energy Technology Laboratory. He said the Energy Department's goal is to get the cost down to $8 a ton, a price at

---

President Bush has endorsed carbon capture and burial as a way to reduce greenhouse gas emissions while promoting energy development.

---

which the emissions could be captured and stored in the United States without increasing the cost to produce electricity by more than 10%.

"It's always going to be cheaper to put carbon dioxide into the air than somewhere else," said Howard Herzog, principal research engineer at the Massachusetts Institute of Technology. "It costs a lot more than anybody seems willing to pay now, but if we decide we really want to solve the climate problem, then it's going to be a cost-effective option."

President Bush has endorsed carbon capture and burial as a way to reduce greenhouse gas emissions while promoting energy development. The Energy Department has a goal for power plants to capture 90% of their carbon emissions by 2012.

Experts caution that sequestration, alone, is only part of the solution.

The United States—which has not signed the Kyoto Protocol—is the world's leading emitter of carbon dioxide, with 1.6 billion tons of emissions annually, about one-quarter of the worldwide total. About 80% of the U.S. emissions come from fossil fuel combustion, according to the Environmental Protection Agency.

Worldwide CO_2 emissions could triple over the next 100 years, reaching 20 billion tons annually, according to the Intergovernmental Panel on Climate Change, the United Nations body that brings together scientists to study global warming.

"We need increased reliance on energy efficiency and increased reliance on renewable energy," said David Hawkins, director of the climate center for the Natural Resources Defense Council, a leading U.S. environmental group. "Carbon storage could provide a third side of the triangle that would allow us to get deep reductions in global warming pollution, but you can't rely on it as the silver bullet."

"We've got to find a way to get industry to get their emissions into the ground instead of into the atmosphere," said Monea, the Weyburn project manager. "One of the most destructive greenhouse gases is carbon dioxide, and mitigation of that greenhouse gas is happening here. We are dealing with this problem, right here, by burying it underground. What's happening here is huge."

California probably has enough capacity in depleted oil fields and subsurface saline deposits to store all the carbon dioxide the state's power plants can produce for the next few centuries, said Benson at the Lawrence Berkeley laboratory. Pilot projects using carbon dioxide injection to enhance oil recovery have been conducted in Kern County, she said.

A consortium of eight partners, including Canada, the United States, the European Union and BP—formerly British Petroleum—have launched a $25-million project to explore new technologies to capture and store carbon gas.

The effort so far has found techniques that reduce costs for geological carbon storage by up to 60%, although more savings are needed before the economics favor doing so on a large scale, said Gardiner Hill, manager of BP's environmental technology group.

"Geological storage is one option that could play a material part in helping us remove emissions and helping the world move forward to a stable amount of carbon dioxide in the atmosphere," Hill said. "The carbon originated from under the ground. We're putting it back."

Carbon Emissions Trading Is New Weapon to Battle Global Warming

By Brad Foss
Associated Press, February 9, 2005

Environmentalists always said there would be a price to pay for all the carbon dioxide being spewed into the atmosphere. Well, now there is.

While prized resources such as oil, gold and wheat have been traded for decades, there is a budding market for one of the industrialized world's abundant but unwanted byproducts: carbon dioxide, a gas produced when fossil fuels are burned and which many scientists believe causes global warming.

If it succeeds, the new market for carbon emissions will reward businesses that minimize their output of this "greenhouse" gas. It will also benefit the environment and thereby prove, advocates say, that making green and being green are compatible goals.

"It's a sign of things to come," said Luis Martinez, an attorney at the Natural Resources Defense Council in New York.

The only mandatory carbon emissions trading program is in Europe. It was created in conjunction with an international treaty on climate change—the Kyoto Protocol—that goes into effect Feb. 16 and caps the amount of carbon dioxide that power plants and fuel-intensive manufacturers in more than two dozen countries are allowed to emit.

A similar program is scheduled to begin in 2008 in Canada, which also signed Kyoto.

By contrast, the United States, one of the few industrialized countries that did not ratify Kyoto, is many years away from compulsory trading or nationwide caps on carbon dioxide, concepts that are strongly opposed by industry and the Bush administration.

However, nine Eastern states are developing a regional cap-and-trade program that will require large power plants from Maine to Delaware to reduce their carbon emissions and California is attempting to place greenhouse gas limits on automakers. Separately, a small group of companies has voluntarily agreed to cap their carbon emissions in the United States as part of an experimental market that is based in Chicago.

"We believe that at some point in the United States there will be mandatory legislation," said Bruce Braine, vice president of strategic policy analysis at American Electric Power Co., a large power producer and one of the founding members of the Chicago Climate Exchange, or CCX. Other members include chemicals giant DuPont Co., computer manufacturer IBM Corp. and electronics maker Motorola Inc.

Under the European Union's Emissions Trading Scheme, some 12,000 industrial plants will be granted a limited number of emissions allowances, or credits, equaling the amount of carbon dioxide they are allowed to emit. Companies that exceed their limits must purchase credits to cover the difference, while those that produce less carbon dioxide than they are legally permitted can sell surplus credits for a profit.

Companies can trade directly with each other or through exchanges located throughout Europe.

By giving the private sector a financial incentive to make their operations more environmentally friendly, proponents believe the market-based approach will accelerate investment in emissions-reduction equipment, create positive reinforcement from investors and spur technological innovation.

"We're confident that once people get used to managing carbon in their businesses it will be successful," said David Hone, climate change adviser at Royal Dutch/Shell Group, which has 46 facilities across Europe that will be regulated under the cap-and-trade system.

Hone said his optimism is based in part on the success of the cap-and-trade system the United States designed more than a decade ago to reduce sulfur dioxide emissions, which cause acid rain. The U.S. sulfur dioxide market, on which the EU's carbon market is based, is widely praised for accelerating emissions reductions at a lower cost than originally anticipated by industry.

But environmentalists and executives said there is much more at stake when it comes to carbon dioxide emissions, both in terms of the ecological benefits and the potential costs to industry.

"Carbon is the mother of all environmental commodities," said Richard Sandor, who helped design the U.S. exchange for sulfur dioxide and is now the chairman of CCX.

The first phase of the EU trading program runs from 2005 through 2007 and the caps will be lowered from one year to the next. While detailed plant-by-plant limits are still being finalized, participants estimate that EU-wide industrial emissions will drop as much as 5 percent by 2008.

The cost to European industry over the next three years is estimated to be a few billion dollars, based on current market prices for carbon dioxide of about 7 euros per ton, according to Ilex Energy Consulting of Oxford, England. Of course, companies with surplus allowances stand to profit an equal amount.

"Frankly, a lot of companies will be hoping that the emissions price is low so they face lower penalties from having to go out and buy allowances," said Andrew Nind, principal consultant at Ilex.

"The main concern of environmentalists is that the governments have been too generous in how many allowances they've given out," Nind said. The lower the price, the less incentive there is to invest in equipment that reduces emissions, he said.

The second phase of the program runs from 2008 through 2012, by which time the European Union must lower its carbon emissions to 8 percent below 1990 levels. Canada must cut its emissions by 6 percent to comply with the Kyoto treaty.

In the United States, carbon-intensive industries successfully lobbied against Kyoto by refuting the threat of global warming itself, and by arguing that the treaty would hurt the global competitiveness of American companies and cause electricity prices to rise.

"There are some notable exceptions, but on the whole it remains an ongoing battle to try to convince corporate America to deal with greenhouse gas emissions," said Ethan Podell of Orbis Energy, which advises companies seeking to devise long-range strategies on carbon emissions.

U.S.–based companies that are already engaged in the issue—either because they have factories in Europe, or are part of CCX—believe early involvement could pay off down the line, in large part due to the logistical expertise they are gaining.

CCX participants agreed beginning in 2003 to cut their carbon emissions by 1 percent per year through 2006, or 4 percent below their baseline, which is determined by their average annual emissions from 1998 through 2001.

There is virtually no chance of federal limits on carbon emissions under the Bush administration, which is openly skeptical of the threat global warming poses. But U.S. executives anticipate a cap-and-trade program for carbon dioxide, at least at the state level, within the next decade.

The Regional Greenhouse Gas Initiative, which aims to produce a preliminary market design by April, already has the support of Connecticut, Delaware, Maine, Massachusetts, New Hampshire, New Jersey, New York, Rhode Island and Vermont.

Global Warming and Nuclear Power

By Richard A. Meserve
Science, January 23, 2004

A recent Massachusetts Institute of Technology (MIT) study on the future of nuclear power argues that nuclear power should be maintained as an energy option because it is an important carbon-free source of power that can potentially make a significant contribution to future electricity supply.* Unfortunately, the study also observes, based on a survey of adults in the United States, that those who are very concerned about global warming are no more likely to support nuclear power than those who are not. Other evidence suggests that the responses in Europe would not be very different. As a result, the MIT authors conclude that public education may be needed to broaden understanding of the links among global warming, fossil fuel usage, and the need for low-carbon energy sources.

For those who are concerned about our future climate, the survey should be disturbing. A realistic response to global warming should involve harnessing a variety of energy options: increased use of renewable energy sources, sequestration of carbon at fossil-fuel plants, enhanced efficiency in energy generation and use, and increased reliance on nuclear power. Because public misunderstanding is likely to manifest itself in the political arena, greater appreciation of the relation between nuclear power and emissions reduction may be essential if use of the nuclear option is to be significantly expanded.

Unfortunately, two institutions that might be expected to explain the facts are largely silent on this issue. Environmental groups include a large and dedicated antinuclear constituency, so even environmentalists who might give nuclear a second look might hesitate to embrace that view publicly. The nuclear industry is reluctant to advance the case because generating companies also rely on fossil-fuel plants (primarily using carbon-intensive coal) for electricity production. This sector thus has a strong disincentive to use global warming as a justification for nuclear power because of the implication of that argument for other components of the companies' supply portfolios.

---

* S. Ansolabehere et al., *The Future of Nuclear Power: An Interdisciplinary MIT Study* (Massachusetts Insititute of Technology, Cambridge, MA, 2003)

As for the Bush administration, it has aggressively supported nuclear power but has carefully avoided emphasizing the link between nuclear power and the global climate's response to increasing concentrations of greenhouse gases. This no doubt reflects the hesitancy that has characterized the administration's approach to the global warming issue.

We thus confront a paradoxical situation. Those who should be the strongest advocates of nuclear power—environmentalists, governmental policy-makers concerned about global warming, and generating companies with an economic stake in nuclear's future—are unable or unwilling to advance the most compelling argument in support of it. Without advocacy by those who see the benefits of nuclear power, it is only to be expected that full exploitation of the nuclear option will be limited or deferred indefinitely.

For those who are serious about confronting global warming, nuclear power should be seen as part of the solution.

Of course, any support for nuclear power should recognize the challenges it presents. Nuclear power is unacceptable unless operators are committed to safe operations and the Nuclear Regulatory Commission exercises careful and detailed oversight. Continuing progress toward the safe final disposition of nuclear waste must be demanded. And tightening safeguards against the diversion of commercial technology to weapons use deserves to be given a high priority around the globe.

Fortunately, all of these challenges can be met. Nuclear power plants have better safety performance today than ever, and future generations of reactors will have design modifications that enhance safety even further. Although debate continues about whether Yucca Mountain is an appropriate disposal site for nuclear waste, the scientific community is in agreement that deep geological disposal somewhere will be a satisfactory means for the disposition of spent fuel. And strengthened international institutions and commitments hold the promise of preventing nuclear power from contributing to the proliferation of nuclear weapons.

For those who are serious about confronting global warming, nuclear power should be seen as part of the solution. Although it is unlikely that many environmental groups will become enthusiastic proponents of nuclear power, the harsh reality is that any serious program to address global warming cannot afford to jettison any technology prematurely. Careful weighing of the risks supports the conclusion that nuclear power at the least must be a bridging technology until other carbon-free energy options become more readily available. The stakes are large, and the scientific and educational community should seek to ensure that the public understands the critical link between nuclear power and climate change.

Is Hydrogen the Solution?

NATURAL RESOURCES DEFENSE COUNCIL (NRDC), APRIL 2004

To address the rapidly accelerating threats posed by global warming and increased dependence on imported oil, the United States and other countries must act to reduce the consumption of petroleum-based fuels in the transportation sector. To manage these threats effectively, we must take near-term actions to significantly increase the efficiency of vehicles that will be sold during the next 25 years. In addition, we must begin now to develop non-petroleum alternatives to fuel the transportation fleets of the future. It will take decades before such new fuels and vehicles will achieve significant market penetration due to the large changes required in vehicle technology and fuel production and distribution systems. We must invest today in two broad areas: incorporating state-of-the-art technology into the vehicles coming on the market in this decade and the next; and developing a new state-of-the-art for vehicles and fuels for the decades that follow.

Among the options for future fuels and fleets, hydrogen and hydrogen fuel cells vehicles have received the most attention and funding to date. The potential of this "hydrogen economy" is drawing increasing attention from Americans concerned about our nation's oil dependence and the threats to our health and well-being from air pollution and global warming. Hydrogen fuel could play a promising long-term role in solving these problems if it is used in high efficiency, non-polluting fuel cells and if it is made from non-polluting energy sources. Hydrogen fuel cells and fuel sources, however, face significant technology, cost, and deployment barriers. A practical assessment of these barriers reveals that it will take at least two decades before hydrogen and fuel cells can begin to make a significant contribution to our energy security, cleaner air, and a safer climate.

The National Academy of Sciences, in a February 2004 report concluded that even the most optimistic predictions for commercializing hydrogen fuel cell vehicles have the first such vehicles reaching commercial showrooms around 2015. It would then take at least another decade or more before hydrogen fuel cell vehicles reach sufficient numbers to begin to have a major impact on the vehicle market, on petroleum imports, and on global warming emissions.[1]

The Academy report concluded: "[H]ydrogen—although it could transform the energy system in the long run—does not represent a short-term solution to any of the nation's energy problems."[2] For this reason, the U.S. Department of Energy "should keep a balanced portfolio of R&D efforts and continue to explore supply-and-demand alternatives that do not depend on hydrogen." The Academy report also stated: "[H]ybrid electric vehicle technology is commercially available today, and benefits from this technology can therefore be utilized immediately."[3] High mileage hybrid electric vehicles, including the Toyota Prius, the Honda Civic Hybrid and hybrid models coming from other automakers, are already competing successfully in the marketplace.

America cannot afford to wait 20 years before we begin to curb our oil dependence and global warming pollution. President Bush touts hydrogen R&D programs with attractive names—"Freedom Car" and "Freedom Fuel"—but these have only long-term pay-off. At the same time, his administration opposes meaningful efforts to reduce oil dependence, air pollution, and global warming *now* with strong

---

America cannot afford to wait 20 years before we begin to curb our oil dependence and global warming pollution.

---

federal performance standards to raise the fuel economy and reduce the global warming pollution of the 17 million conventional cars and trucks that will be sold each year for the next 20 years. In fact, the administration just recently extended alternative fuel mileage credits that actually have the affect of increasing petroleum consumption and more than offsetting the modest 1.5 mpg increase in light truck fuel economy that the administration recently adopted.[4]

Strong federal standards would motivate automakers to commercialize the cleaner, more efficient technologies that are already available today—more fuel-efficient engines, transmissions, and other components for conventional gasoline-powered cars, as well as more hybrid gas-electric vehicles. Strong standards can also help accelerate a transition to a clean and sustainable fuel supply.

For more than 30 years, California has played a special role in national efforts to clean up our cars and trucks. Each generation of the pollution control technology that is now found on every vehicle nationwide was pioneered in California under the state's own clean air standards. Now California is playing a similar role in developing the hydrogen economy. California's Zero Emission Vehicle (ZEV) program is spurring the early introduction of critical technologies, such as electric drives and fuel cells, into our smoggiest cities. But what is lacking are meaningful federal standards that go beyond long-term R&D to make cleaner vehicles a reality nationwide.

Other countries, notably various EU countries, are successfully pursuing a portfolio of near- and long-term options for improving energy security and reducing environmental impacts from the transportation sector. In addition, China is investing heavily in research and development of fuel cell vehicles and hydrogen production and storage options while also developing new fuel economy standards for conventional vehicles that are tougher on gas-guzzling SUVs than those in the United States.

What is the "hydrogen economy"?

America must immediately begin kicking the petroleum habit and substantially reducing our global warming pollution. Hydrogen is one promising long-term successor to the petroleum fuels that currently power 97 percent of our vehicles.

It really matters, however, where the hydrogen comes from. It can be produced from a wide variety of sources: fossil fuels, biomass, or electrolysis of water, where the electricity used for electrolysis can also be generated from a range of sources. One of hydrogen's primary advantages is that it can be produced from a

---

Hydrogen is one promising long-term successor to the petroleum fuels that currently power 97 percent of our vehicles.

---

diverse number of entirely domestic and renewable sources. Hydrogen can increase our energy security if it is made from the former. It can improve our environment if it is made from the latter (e.g., biofuels or electrolysis powered by wind or solar electricity) or if the carbon pollution from fossil fuel sources is successfully pumped back underground for permanent storage.

It also matters how hydrogen is used, whether in fuel cells or conventional internal combustion engines. Fuel cells are inherently more efficient than gasoline engines. They convert hydrogen to electricity efficiently and without high temperature combustion, through a chemical reaction, much as a standard battery does. In this reaction, the hydrogen fuel reacts with oxygen from air to produce electricity, and the only emission from the vehicle is water vapor. There is no smog-forming or global warming pollution from the vehicle.

Tremendous progress has been made over the last decade in developing fuel cell vehicle prototypes, spurred in large part by the California Zero Emission Vehicle program. Every major automobile manufacturer has a fuel cell vehicle (FCV) prototype and a small number of FCVs are on the roads in California, Washington, D.C., Japan, and Europe.

While these investments have spurred many technological improvements, they have also revealed the many difficulties of designing cost-effective and safe fuel cells and hydrogen storage and dispensing systems. Before FCVs will be ready for mass production, significant technical and cost challenges must be resolved. These include the high current cost of fuel cells, their lack of durability, limited driving range due to limitations in on-board fuel storage technology, issues of safety and lack of infrastructure to produce and distribute the hydrogen to large numbers of consumers.

These considerations only underline the importance of taking concrete steps now with technology that is already available in the near-term, and the danger of putting all our eggs in the hydrogen basket even for the long term. Building a sustainable transportation energy system must include strong clean air and fuel economy standards in the near-term and ultimately, a more balanced portfolio of R&D focusing on other clean fuel technologies as well as hydrogen in the long-term. This combined approach will begin immediately reducing our oil imports and global warming pollution with technologies available today, while exploring and preserving a range of potential future technology options.

Other vehicle and fuel options appear to have at least as much potential as hydrogen to substantially reduce oil consumption and global warming pollution from vehicles. Ethanol used in high efficiency vehicles, such as hybrids, is one such option if the ethanol is made from agricultural waste or energy crops (cellulosic ethanol) instead of today's corn-based ethanol. Furthermore, a breakthrough in the efficiency of chemical batteries or other technologies for storing electricity could allow rechargeable electric vehicles to penetrate the mass market.

Why will it take at least two decades for hydrogen fuel cells to contribute to reducing our petroleum dependence and global warming emissions?

Currently, there are 180 million passenger vehicles in the U.S., virtually all of which run on gasoline or diesel. Even after hydrogen fuel cell vehicles make it to the showroom, it will take many decades before the majority of vehicles on the road are fuel cell–powered.

Steady improvement to today's gasoline vehicles would have a much bigger near-term and mid-term impact than even the most aggressive deployment of FCVs. Today's fuel-saving gasoline technologies can be deployed now and can run off the current fueling infrastructure. FCVs will require an entirely new infrastructure that faces significant technical and economic barriers.

As shown in NRDC's recent *Dangerous Addiction* report,[5] increased fuel economy could save almost 25 times more oil between now and 2020 than FCVs. By 2030, when fuel cells could be more prevalent, oil savings from conventional and hybrid vehicle fuel-economy improvements are still five times as great as those from fuel cell vehicles.

To illustrate this point, we compared the oil savings for a fuel-economy proposal of 40 miles per gallon by 2012 and 55 miles per gallon by 2020 with projected savings from a fuel-cell target of 100,000 fuel-cell vehicles per year by 2010 and 2.5 million per year in 2020 *without* any improvement in the fuel economy of conventional vehicles. Figure 1 shows that oil savings up to 2030 from fuel-cell technology—even on an optimistic timeline—are dwarfed by the gains that can be achieved by raising the gas mileage of conventional American cars and trucks.

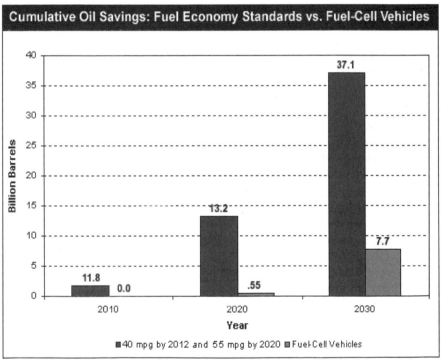

Cumulative Oil Savings: Fuel Economy Standards vs. Fuel-Cell Vehicles

Fuel economy scenarios from *Dangerous Addiction: Ending America's Oil Dependence*, NRDC & UCS, January 2002.

All hydrogen is *not* created equal

Hydrogen can be extracted from a variety of sources, including natural gas, water, biomass, and even oil or coal. Currently, most hydrogen is extracted from natural gas, oil, and coal, while releasing carbon dioxide from those fuels into the atmosphere. Small amounts of hydrogen are produced by electrolysis (splitting water into hydrogen and oxygen with electricity) using conventional (mainly fossil-fueled) sources of electricity, which also results in carbon dioxide emissions. Hydrogen can also be made from renewable sources, and from fossil fueled plants that pump carbon dioxide underground. These sources have very different environmental and energy security implications.[6]

Natural Gas Steam Methane Reforming. At present, natural gas is the leading "feedstock" under consideration for near-term hydrogen production in the U.S. Natural gas is a non-renewable resource, and hydrogen production from reforming natural gas would result in substantial carbon dioxide emissions. The greatest supplies of natural gas are found in ecologically sensitive locations and in politically unstable parts of the world, as is petroleum. To power 40% of the fleet with hydrogen from natural gas in 2025, even using higher efficiency fuel cells, would require a 33% increase in natural gas supply from projected 2025 levels.[7] Since natural gas is already in heavy demand as the cleanest fossil fuel for power plants, alternative sources of hydrogen production are needed. Also, unless global warming emissions are removed and stored underground, the natural gas process will continue to contribute to global warming. In other words, while natural gas might work as a transition fuel, a hydrogen economy based on natural gas would be neither sustainable nor secure in the long term.

Electrolysis and Renewables. Electrolysis using renewably generated electricity, such as wind or photovoltaics, would produce a domestic, non-polluting hydrogen transportation fuel. Unfortunately, hydrogen produced today by this method can be more than 3 times the cost of an equivalent gallon of gasoline. On the other hand, if the electricity is supplied from the present-day electrical grid, currently more than 50% coal-fired, while somewhat less expensive than hydrogen production from renewable-based electrolysis it would generate even larger amounts of carbon emissions than the natural gas process. In order for hydrogen fuel cell vehicles to reduce global warming pollution, the electrolysis process will have to become more efficient, and the electricity driving it will need to be produced from a high percentage of low- to zero-carbon sources (e.g., renewables or coal with carbon capture and storage). Under current projections, electrolytic hydrogen from grid electricity in the U.S. would likely create a net increase in global warming emissions for at least the next couple of decades.[8, 9]

Biomass and Biofuels. Dedicated sustainable energy crops could also serve as a crucial part of a hydrogen economy since they can serve directly as a carbon-free source of hydrogen through biomass gasification, or be converted to cellulosic-based ethanol as an intermediate step and then to hydrogen. This later option is attractive since ethanol as a room temperature liquid fuel is substantially easier to transport. Energy crops could also diversify agricultural markets, help stabilize the agricultural economy, contribute to rural economic development, and reduce the adverse impacts of agricultural subsidies on developing countries. More research and development of the production processes of biomass to hydrogen and ethanol-to-hydrogen is needed to make this source of energy a cost-effective and viable option.

Eventually, hydrogen could also be produced directly from renewable sources through photoelectrochemical or photobiological processes, but these are still at an early stage of research and development and could take decades to come to fruition.

Coal Gasification. The Bush administration's hydrogen fuel initiative has strongly emphasized producing hydrogen from coal. While coal has the advantage of being a domestic resource, its drawbacks include high emissions of carbon dioxide and other pollutants and land and water impacts from current mining practices. If these environmental problems are addressed through the use of technologies such as coal gasification with carbon capture and storage[10] and low environmental impact mining, then hydrogen from coal could potentially become a sustainable source of hydrogen over the long term, both in the U.S. and other coal-rich countries. But for these technologies to come to market, the U.S. must establish enforceable limits on carbon dioxide emissions and demonstrate the viability of large-scale carbon storage projects as soon as possible. The Department of Energy is exploring initiatives that gasify coal producing both hydrogen and electricity while storing the waste carbon dioxide in a geologic formation.

Carbon storage needs further research and development. But it is not the "silver bullet" solution for producing clean hydrogen as portrayed by the Bush administration. Carbon storage, if proven to be safe, permanent, and environmentally benign, should be only one component of a portfolio of options for producing hydrogen. The emphasis should be on producing hydrogen from intrinsically clean sources, such as wind, solar, and sustainable biomass crops. Yet to date the administration has failed to place a high enough priority on these clean options. The administration also supports nuclear power, which could produce hydrogen with no global warming emissions. But expanding nuclear power presents significant safety, waste disposal, security, and land use risks. A portfolio of truly clean, renewable, and domestic sources of hydrogen must ultimately form the basis for any future hydrogen economy.

Steering a New Course

Hydrogen is not necessarily going to be a magic solution to America's global warming pollution and our country's petroleum dependence. Nevertheless, placed in perspective hydrogen has the potential to play a major role and therefore must be explored aggressively. However, these efforts must be accompanied by near-term actions. First and foremost, we need policies that mobilize the technologies we already have. That means raising fuel economy standards, accelerating hybrid vehicle production, and actively developing options such as cellulosic ethanol. To complement these actions we need a realistic, diverse and aggressive R&D effort to develop the longer-term technologies, such as hydrogen, that hold significant promise.

Numerous analyses demonstrate that if we do nothing to curb global warming or cut our foreign oil dependence until a hydrogen economy is ready, these problems will be too big to solve. A responsible, economically sensible energy policy that effectively addresses both global warming and energy security must include:

- Aggressive near-term performance standards requiring automakers to cut the oil consumption and global warming pollution of their fleets by deploying technologies that reduce carbon emissions, save fuel, or do both.

- Accelerated deployment of off-the-shelf and near-term vehicle technologies and fuels that can reduce oil dependence and global warming pollution while providing a potential transition to hydrogen fuel cells.

- Strategic deployment of fuel cells. In the coming decade, we need to conduct fleet testing of FCVs in our smoggiest cities, in order to get real world experience with the commercialization challenges while cleaning up the air.

- Continued research and development into fuel cell vehicle technology and advanced hydrogen system technologies, complemented with research, development, and deployment of the capacity to make hydrogen from wind, solar power, and biofuels, and from fossil-fuel sources without carbon emissions. These are the most sustainable energy sources for potential future hydrogen production.

- Long-term research and development efforts that continue to explore alternative clean energy options to hydrogen such as biofuels or battery electric vehicles.

Notes

1. National Research Council of the National Academy of Sciences, "The Hydrogen Economy: Opportunities, Costs, Barriers, and R&D Needs," February 2004.

2. Ibid, p. 6–14.

3. Ibid, p. ES–2.

4. Extending the alternative fuel credits is estimated to increase the nation's petroleum consumption by around 3 billion gallons from 2005 through 2008 (see Dept. of Transportation, Environmental Protection Agency, Dept. of Energy, Report to Congress: Effects of the Alternative Motor Fuels Act CAFE Incentives Policy (Mar. 2002)), while the 1.5 mpg increase in light truck standards is projected to save just 0.9 billion gallons over the same period (see Dept. of Transportation, National Highway Traffic Safety Administration, Light Truck Average Fuel Economy Standards Model Years 2005–2007, Notice of Final Rule, 68 Fed. Reg. 16868, 16898 (April 7, 2003)).

5. Daniel Lashof and Roland Hwang, *Dangerous Addiction 2003—Breaking the Chain of Oil Dependence*, March 2003.

6. Wang, et al. "Well-to-Wheels Energy and Emission Impacts of Vehicle/Fuel Systems Development and Applications of the GREET Model," 2003; National

Research Council of the National Academy of Sciences, "The Hydrogen Economy: Opportunities, Costs, Barriers, and R&D Needs," February 2004.

7. Joseph Romm, Hydrogen and Fuel Cells: A Technology and Policy Overview, prepared for the National Commission on Energy Policy by The Center for Energy and Climate Solutions, October, 2003.

8. C. E. (Sandy) Thomas, John P. Reardon, Franklin D. Lomax, Jr., Jennifer Pinyan & Ira F. Kuhn, Jr., Proceedings of the 2001 DOE Hydrogen Program Review, "Distributes Hydrogen Fueling Systems Analysis."

9. Today, the U.S. generates about 71 percent of its electricity from fossil fuels. The Department of Energy projects that this share will increase to 77 percent by 2025 while the percentage of renewable generated electricity will stay constant at about 7–8 percent, US DOE Annual Energy Outlook 2004.

10. Carbon storage is the process of permanently storing carbon dioxide into geologic or ocean reservoirs, and if successful could be used to reduce these emissions from burning coal and other fossil fuels, making them more acceptable sources of hydrogen, or electricity production.

The Warming Is Global but the Legislating, in the U.S., Is All Local

By Jennifer 8. Lee
The New York Times, October 29, 2003

Motivated by environmental and economic concerns, states have become the driving force in efforts to combat global warming even as mandatory programs on the federal level have largely stalled.

At least half of the states are addressing global warming, whether through legislation, lawsuits against the Bush administration or programs initiated by governors.

In the last three years, state legislatures have passed at least 29 bills, usually with bipartisan support. The most contentious is California's 2002 law to set strict limits for new cars on emissions of carbon dioxide, the gas that scientists say has the greatest role in global warming.

While few of the state laws will have as much impact as California's, they are not merely symbolic. In addition to caps on emissions of gases like carbon dioxide that can cause the atmosphere to heat up like a greenhouse, they include registries to track such emissions, efforts to diversify fuel sources and the use of crops to capture carbon dioxide by taking it out of the atmosphere and into the ground.

Aside from their practical effects, supporters say, these efforts will put pressure on Congress and the administration to enact federal legislation, if only to bring order to a patchwork of state laws.

States are moving ahead in large part to fill the vacuum that has been left by the federal government, said David Danner, the energy adviser for Gov. Gary Locke of Washington.

"We hope to see the problem addressed at the federal level," Mr. Danner said, "but we're not waiting around."

There are some initiatives in Congress, but for the moment even their backers acknowledge that they are doomed, given strong opposition from industry, the Bush administration—which favors voluntary controls—and most Congressional Republicans.

This week, the Senate is scheduled to vote on a proposal to create a national regulatory structure for carbon dioxide. This would be the first vote for either house on a measure to restrict the gas.

The proposal's primary sponsors, Senator John McCain, Republican of Arizona, and Senator Joseph I. Lieberman, Democrat of Connecticut, see it mainly as a way to force senators to take a position on the issue, given the measure's slim prospects.

States are acting partly because of predictions that global warming could damage local economies by harming agriculture, eroding shorelines and hurting tourism.

"We're already seeing things which may be linked to global warming here in the state," Mr. Danner said. "We have low snowpack, increased forest fire danger."

Environmental groups and officials in state governments say that energy initiatives are easier to move forward on the local level because they span constituencies—industrial and service sectors, Democrat and Republican, urban and rural.

While the coal, oil and automobile industries have big lobbies in Washington, the industry presence is diluted on the state level. Environmental groups say this was crucial to winning a legislative battle over automobile emissions in California, where the automo-

---

> *States are acting partly because of predictions that global warming could damage local economies.*

---

bile industry did not have a long history of large campaign donations and instead had to rely on a six-month advertising campaign to make its case.

Local businesses are also interested in policy decisions because of concerns about long-term energy costs, said Christopher James, director of air planning and standards for the Connecticut Department of Environmental Protection. As a result, environmental groups are shifting their efforts to focus outside Washington.

Five years ago the assumption was that the climate treaty known as the Kyoto Protocol was the only effort in town, said Rhys Roth, the executive director of Climate Solutions, which works on global warming issues in the Pacific Northwest states. But since President Bush rejected the Kyoto pact in 2001, local groups have been emerging on the regional, state and municipal levels.

The Climate Action Network, a worldwide conglomeration of nongovernment organizations working on global warming, doubled its membership of state and local groups in the last two years.

The burst of activity is not limited to the states with a traditional environmental bent.

At least 15 states, including Texas and Nevada, are forcing their state electric utilities to diversify beyond coal and oil to energy sources like wind and solar power.

Even rural states are linking their agricultural practices to global warming. Nebraska, Oklahoma and Wyoming have all passed initiatives in anticipation of future greenhouse-gas emission trading, hoping they can capitalize on their forests and crops to capture carbon dioxide during photosynthesis.

Cities are also adopting new energy policies. San Franciscans approved a $100 million bond initiative in 2001 to pay for solar panels for municipal buildings, including the San Francisco convention center.

The rising level of state activity is causing concern among those who oppose carbon dioxide regulation.

"I believe the states are being used to force a federal mandate," said Sandy Liddy Bourne, who does research on global warming for the American Legislative Exchange Council, a group contending that carbon dioxide should not be regulated because it is not a pollutant. "Rarely do you see so many bills in one subject area introduced across the country."

The council started tracking state legislation, which they call son-of-Kyoto bills, weekly after they noticed a significant rise in greenhouse-gas-related legislation two years ago. This year, the council says, 24 states have introduced 90 bills that would build frameworks for regulating carbon dioxide. Sixty-six such bills were introduced in all of 2001 and 2002.

Some of the activity has graduated to a regional level. Last summer, Gov. George E. Pataki of New York invited 10 Northeastern states to set up a regional trading network where power plants could buy and sell carbon dioxide credits in an effort to lower overall emissions. In 2001, six New England states entered into an agreement with Canadian provinces to cap overall emissions by 2010. Last month, California, Washington and Oregon announced that they would start looking at shared strategies to address global warming.

To be sure, some states have decided not to embrace policies to combat global warming. Six—Alabama, Illinois, Kentucky, Oklahoma, West Virginia and Wyoming—have explicitly passed laws against any mandatory reductions in greenhouse gas emissions.

"My concern," said Ms. Bourne, "is that members of industry and environment groups will go to the federal government to say: 'There is a patchwork quilt of greenhouse-gas regulations across the country. We cannot deal with the 50 monkeys. We must have one 800-pound gorilla. Please give us a federal mandate.'" Indeed, some environmentalists say this is precisely their strategy.

States developed their own air toxics pollution programs in the 1980's, which resulted in different regulations and standards across the country. Industry groups, including the American Chemistry Council, eventually lobbied Congress for federal standards, which were incorporated into the 1990 Clean Air Act amendments.

A number of states are trying to compel the federal government to move sooner rather than later. On Thursday, 12 states, including New York, with its Republican governor, and three cities sued the Environmental Protection Agency for its recent decision not to regulate greenhouse-gas pollutants under the Clean Air Act, a reversal of the agency's previous stance under the Clinton administration.

"Global warming cannot be solely addressed at the state level," said Tom Reilly, the Massachusetts attorney general. "It's a problem that requires a federal approach."

Appendix

Kyoto Protocol to the United Nations Framework Convention on Climate Change

The Parties to this Protocol,
Being Parties to the United Nations Framework Convention on Climate Change, hereinafter referred to as "the Convention,"
In pursuit of the ultimate objective of the Convention as stated in its Article 2,
Recalling the provisions of the Convention,
Being guided by Article 3 of the Convention,
Pursuant to the Berlin Mandate adopted by decision 1/CP.1 of the
Conference of the Parties to the Convention at its first session,
Have agreed as follows:

Article 1

For the purposes of this Protocol, the definitions contained in Article 1 of the Convention shall apply. In addition:

1. "Conference of the Parties" means the Conference of the Parties to the Convention.
2. "Convention" means the United Nations Framework Convention on Climate Change, adopted in New York on 9 May 1992.
3. "Intergovernmental Panel on Climate Change" means the Intergovernmental Panel on Climate Change established in 1988 jointly by the World Meteorological Organization and the United Nations Environment Programme.
4. "Montreal Protocol" means the Montreal Protocol on Substances that Deplete the Ozone Layer, adopted in Montreal on 16 September 1987 and as subsequently adjusted and amended.
5. "Parties present and voting" means Parties present and casting an affirmative or negative vote.
6. "Party" means, unless the context otherwise indicates, a Party to this Protocol.
7. "Party included in Annex I" means a Party included in Annex I to the Convention, as may be amended, or a Party which has made a notification under Article 4, paragraph 2(g), of the Convention.

Article 2

1. Each Party included in Annex I, in achieving its quantified emission limitation and reduction commitments under Article 3, in order to promote sustainable development, shall:

   (a) Implement and/or further elaborate policies and measures in accordance with its national circumstances, such as:

      (i) Enhancement of energy efficiency in relevant sectors of the national economy;

      (ii) Protection and enhancement of sinks and reservoirs of greenhouse gases not controlled by the Montreal Protocol, taking into account its commitments under relevant international environmental agreements; promotion of sustainable forest management practices, afforestation and reforestation;

      (iii) Promotion of sustainable forms of agriculture in light of climate change considerations;

      (iv) Research on, and promotion, development and increased use of, new and renewable forms of energy, of carbon dioxide sequestration technologies and of advanced and innovative environmentally sound technologies;

      (v) Progressive reduction or phasing out of market imperfections, fiscal incentives, tax and duty exemptions and subsidies in all greenhouse gas emitting sectors that run counter to the objective of the Convention and application of market instruments;

      (vi) Encouragement of appropriate reforms in relevant sectors aimed at promoting policies and measures which limit or reduce emissions of greenhouse gases not controlled by the Montreal Protocol;

(vii) Measures to limit and/or reduce emissions of greenhouse gases not controlled by the Montreal Protocol in the transport sector;

(viii) Limitation and/or reduction of methane emissions through recovery and use in waste management, as well as in the production, transport and distribution of energy;

(b) Cooperate with other such Parties to enhance the individual and combined effectiveness of their policies and measures adopted under this Article, pursuant to Article 4, paragraph 2(e)(i), of the Convention. To this end, these Parties shall take steps to share their experience and exchange information on such policies and measures, including developing ways of improving their comparability, transparency and effectiveness. The Conference of the Parties serving as the meeting of the Parties to this Protocol shall, at its first session or as soon as practicable thereafter, consider ways to facilitate such cooperation, taking into account all relevant information.

2. The Parties included in Annex I shall pursue limitation or reduction of emissions of greenhouse gases not controlled by the Montreal Protocol from aviation and marine bunker fuels, working through the International Civil Aviation Organization and the International Maritime Organization, respectively.

3. The Parties included in Annex I shall strive to implement policies and measures under this Article in such a way as to minimize adverse effects, including the adverse effects of climate change, effects on international trade, and social, environmental and economic impacts on other Parties, especially developing country Parties and in particular those identified in Article 4, paragraphs 8 and 9, of the Convention, taking into account Article 3 of the Convention. The Conference of the Parties serving as the meeting of the Parties to this Protocol may take further action, as appropriate, to promote the implementation of the provisions of this paragraph.

4. The Conference of the Parties serving as the meeting of the Parties to this Protocol, if it decides that it would be beneficial to coordinate any of the policies and measures in paragraph 1(a) above, taking into account different national circumstances and potential effects, shall consider ways and means to elaborate the coordination of such policies and measures.

Article 3

1. The Parties included in Annex I shall, individually or jointly, ensure that their aggregate anthropogenic carbon dioxide equivalent emissions of the greenhouse gases listed in Annex A do not exceed their assigned amounts, calculated pursuant to their quantified emission limitation and reduction commitments inscribed in Annex B and in accordance with the provisions of this Article, with a view to reducing their overall emissions of such gases by at least 5 per cent below 1990 levels in the commitment period 2008 to 2012.

2. Each Party included in Annex I shall, by 2005, have made demonstrable progress in achieving its commitments under this Protocol.

3. The net changes in greenhouse gas emissions by sources and removals by sinks resulting from direct human-induced land-use change and forestry activities, limited to afforestation, reforestation and deforestation since 1990, measured as verifiable changes in carbon stocks in each commitment period, shall be used to meet the commitments under this Article of each Party included in Annex I. The greenhouse gas emissions by sources and removals by sinks associated with those activities shall be reported in a transparent and verifiable manner and reviewed in accordance with Articles 7 and 8.

4. Prior to the first session of the Conference of the Parties serving as the meeting of the Parties to this Protocol, each Party included in Annex I shall provide, for consideration by the Subsidiary Body for Scientific and Technological Advice, data to establish its level of carbon stocks in 1990 and to enable an estimate to be made of its changes in carbon stocks in subsequent years. The Conference of the Parties serving as the meeting of the Parties to this Protocol shall, at its first session or as soon as practicable thereafter, decide upon modalities, rules and guidelines as to how, and which, additional human-induced activities related to changes in greenhouse gas emissions by sources and removals by sinks in the agricultural

soils and the land-use change and forestry categories shall be added to, or subtracted from, the assigned amounts for Parties included in Annex I, taking into account uncertainties, transparency in reporting, verifiability, the methodological work of the Intergovernmental Panel on Climate Change, the advice provided by the Subsidiary Body for Scientific and Technological Advice in accordance with Article 5 and the decisions of the Conference of the Parties. Such a decision shall apply in the second and subsequent commitment periods. A Party may choose to apply such a decision on these additional human-induced activities for its first commitment period, provided that these activities have taken place since 1990.

5. The Parties included in Annex I undergoing the process of transition to a market economy whose base year or period was established pursuant to decision 9/CP.2 of the Conference of the Parties at its second session shall use that base year or period for the implementation of their commitments under this Article. Any other Party included in Annex I undergoing the process of transition to a market economy which has not yet submitted its first national communication under Article 12 of the Convention may also notify the Conference of the Parties serving as the meeting of the Parties to this Protocol that it intends to use an historical base year or period other than 1990 for the implementation of its commitments under this Article. The Conference of the Parties serving as the meeting of the Parties to this Protocol shall decide on the acceptance of such notification.

6. Taking into account Article 4, paragraph 6, of the Convention, in the implementation of their commitments under this Protocol other than those under this Article, a certain degree of flexibility shall be allowed by the Conference of the Parties serving as the meeting of the Parties to this Protocol to the Parties included in Annex I undergoing the process of transition to a market economy.

7. In the first quantified emission limitation and reduction commitment period, from 2008 to 2012, the assigned amount for each Party included in Annex I shall be equal to the percentage inscribed for it in Annex B of its aggregate anthropogenic carbon dioxide equivalent emissions of the greenhouse gases listed in Annex A in 1990, or the base year or period determined in accordance with paragraph 5 above, multiplied by five. Those Parties included in Annex I for whom land-use change and forestry constituted a net source of greenhouse gas emissions in 1990 shall include in their 1990 emissions base year or period the aggregate anthropogenic carbon dioxide equivalent emissions by sources minus removals by sinks in 1990 from land-use change for the purposes of calculating their assigned amount.

8. Any Party included in Annex I may use 1995 as its base year for hydrofluorocarbons, perfluorocarbons and sulphur hexafluoride, for the purposes of the calculation referred to in paragraph 7 above.

9. Commitments for subsequent periods for Parties included in Annex I shall be established in amendments to Annex B to this Protocol, which shall be adopted in accordance with the provisions of Article 21, paragraph 7. The Conference of the Parties serving as the meeting of the Parties to this Protocol shall initiate the consideration of such commitments at least seven years before the end of the first commitment period referred to in paragraph 1 above.

10. Any emission reduction units, or any part of an assigned amount, which a Party acquires from another Party in accordance with the provisions of Article 6 or of Article 17 shall be added to the assigned amount for the acquiring Party.

11. Any emission reduction units, or any part of an assigned amount, which a Party transfers to another Party in accordance with the provisions of Article 6 or of Article 17 shall be subtracted from the assigned amount for the transferring Party.

12. Any certified emission reductions which a Party acquires from another Party in accordance with the provisions of Article 12 shall be added to the assigned amount for the acquiring Party.

13. If the emissions of a Party included in Annex I in a commitment period are less than its assigned amount under this Article, this difference shall, on request of that Party, be added to the assigned amount for that Party for subsequent commitment periods.

14. Each Party included in Annex I shall strive to implement the commitments mentioned in paragraph 1 above in such a way as to minimize adverse social, environmental and economic

impacts on developing country Parties, particularly those identified in Article 4, paragraphs 8 and 9, of the Convention. In line with relevant decisions of the Conference of the Parties on the implementation of those paragraphs, the Conference of the Parties serving as the meeting of the Parties to this Protocol shall, at its first session, consider what actions are necessary to minimize the adverse effects of climate change and/or the impacts of response measures on Parties referred to in those paragraphs. Among the issues to be considered shall be the establishment of funding, insurance and transfer of technology.

Article 4

1. Any Parties included in Annex I that have reached an agreement to fulfil their commitments under Article 3 jointly, shall be deemed to have met those commitments provided that their total combined aggregate anthropogenic carbon dioxide equivalent emissions of the greenhouse gases listed in Annex A do not exceed their assigned amounts calculated pursuant to their quantified emission limitation and reduction commitments inscribed in Annex B and in accordance with the provisions of Article 3. The respective emission level allocated to each of the Parties to the agreement shall be set out in that agreement.

2. The Parties to any such agreement shall notify the secretariat of the terms of the agreement on the date of deposit of their instruments of ratification, acceptance or approval of this Protocol, or accession thereto. The secretariat shall in turn inform the Parties and signatories to the Convention of the terms of the agreement.

3. Any such agreement shall remain in operation for the duration of the commitment period specified in Article 3, paragraph 7.

4. If Parties acting jointly do so in the framework of, and together with, a regional economic integration organization, any alteration in the composition of the organization after adoption of this Protocol shall not affect existing commitments under this Protocol. Any alteration in the composition of the organization shall only apply for the purposes of those commitments under Article 3 that are adopted subsequent to that alteration.

5. In the event of failure by the Parties to such an agreement to achieve their total combined level of emission reductions, each Party to that agreement shall be responsible for its own level of emissions set out in the agreement.

6. If Parties acting jointly do so in the framework of, and together with, a regional economic integration organization which is itself a Party to this Protocol, each member State of that regional economic integration organization individually, and together with the regional economic integration organization acting in accordance with Article 24, shall, in the event of failure to achieve the total combined level of emission reductions, be responsible for its level of emissions as notified in accordance with this Article.

Article 5

1. 1. Each Party included in Annex I shall have in place, no later than one year prior to the start of the first commitment period, a national system for the estimation of anthropogenic emissions by sources and removals by sinks of all greenhouse gases not controlled by the Montreal Protocol. Guidelines for such national systems, which shall incorporate the methodologies specified in paragraph 2 below, shall be decided upon by the Conference of the Parties serving as the meeting of the Parties to this Protocol at its first session.

2. Methodologies for estimating anthropogenic emissions by sources and removals by sinks of all greenhouse gases not controlled by the Montreal Protocol shall be those accepted by the Intergovernmental Panel on Climate Change and agreed upon by the Conference of the Parties at its third session. Where such methodologies are not used, appropriate adjustments shall be applied according to methodologies agreed upon by the Conference of the Parties serving as the meeting of the Parties to this Protocol at its first session. Based on the work of, *inter alia*, the Intergovernmental Panel on Climate Change and advice provided by the Subsidiary Body for Scientific and Technological Advice, the Conference of the Parties serving as the meeting of the Parties to this Protocol shall regularly review and, as appropriate, revise such methodologies and adjustments, taking fully into account any relevant decisions by the Conference of the Parties. Any revision to methodologies or adjustments shall be used

only for the purposes of ascertaining compliance with commitments under Article 3 in respect of any commitment period adopted subsequent to that revision.

3. The global warming potentials used to calculate the carbon dioxide equivalence of anthropogenic emissions by sources and removals by sinks of greenhouse gases listed in Annex A shall be those accepted by the Intergovernmental Panel on Climate Change and agreed upon by the Conference of the Parties at its third session. Based on the work of, *inter alia*, the Intergovernmental Panel on Climate Change and advice provided by the Subsidiary Body for Scientific and Technological Advice, the Conference of the Parties serving as the meeting of the Parties to this Protocol shall regularly review and, as appropriate, revise the global warming potential of each such greenhouse gas, taking fully into account any relevant decisions by the Conference of the Parties. Any revision to a global warming potential shall apply only to commitments under Article 3 in respect of any commitment period adopted subsequent to that revision.

Article 6

1. For the purpose of meeting its commitments under Article 3, any Party included in Annex I may transfer to, or acquire from, any other such Party emission reduction units resulting from projects aimed at reducing anthropogenic emissions by sources or enhancing anthropogenic removals by sinks of greenhouse gases in any sector of the economy, provided that:

   (a) Any such project has the approval of the Parties involved;

   (b) Any such project provides a reduction in emissions by sources, or an enhancement of removals by sinks, that is additional to any that would otherwise occur;

   (c) It does not acquire any emission reduction units if it is not in compliance with its obligations under Articles 5 and 7; and

   (d) The acquisition of emission reduction units shall be supplemental to domestic actions for the purposes of meeting commitments under Article 3.

2. The Conference of the Parties serving as the meeting of the Parties to this Protocol may, at its first session or as soon as practicable thereafter, further elaborate guidelines for the implementation of this Article, including for verification and reporting.

3. A Party included in Annex I may authorize legal entities to participate, under its responsibility, in actions leading to the generation, transfer or acquisition under this Article of emission reduction units.

4. If a question of implementation by a Party included in Annex I of the requirements referred to in this Article is identified in accordance with the relevant provisions of Article 8, transfers and acquisitions of emission reduction units may continue to be made after the question has been identified, provided that any such units may not be used by a Party to meet its commitments under Article 3 until any issue of compliance is resolved.

Article 7

1. Each Party included in Annex I shall incorporate in its annual inventory of anthropogenic emissions by sources and removals by sinks of greenhouse gases not controlled by the Montreal Protocol, submitted in accordance with the relevant decisions of the Conference of the Parties, the necessary supplementary information for the purposes of ensuring compliance with Article 3, to be determined in accordance with paragraph 4 below.

2. Each Party included in Annex I shall incorporate in its national communication, submitted under Article 12 of the Convention, the supplementary information necessary to demonstrate compliance with its commitments under this Protocol, to be determined in accordance with paragraph 4 below.

3. Each Party included in Annex I shall submit the information required under paragraph 1 above annually, beginning with the first inventory due under the Convention for the first year of the commitment period after this Protocol has entered into force for that Party. Each such Party shall submit the information required under paragraph 2 above as part of the first national communication due under the Convention after this Protocol has entered into force for it and after the adoption of guidelines as provided for in paragraph 4 below. The

frequency of subsequent submission of information required under this Article shall be determined by the Conference of the Parties serving as the meeting of the Parties to this Protocol, taking into account any timetable for the submission of national communications decided upon by the Conference of the Parties.

4. The Conference of the Parties serving as the meeting of the Parties to this Protocol shall adopt at its first session, and review periodically thereafter, guidelines for the preparation of the information required under this Article, taking into account guidelines for the preparation of national communications by Parties included in Annex I adopted by the Conference of the Parties. The Conference of the Parties serving as the meeting of the Parties to this Protocol shall also, prior to the first commitment period, decide upon modalities for the accounting of assigned amounts.

Article 8

1. The information submitted under Article 7 by each Party included in Annex I shall be reviewed by expert review teams pursuant to the relevant decisions of the Conference of the Parties and in accordance with guidelines adopted for this purpose by the Conference of the Parties serving as the meeting of the Parties to this Protocol under paragraph 4 below. The information submitted under Article 7, paragraph 1, by each Party included in Annex I shall be reviewed as part of the annual compilation and accounting of emissions inventories and assigned amounts. Additionally, the information submitted under Article 7, paragraph 2, by each Party included in Annex I shall be reviewed as part of the review of communications.

2. Expert review teams shall be coordinated by the secretariat and shall be composed of experts selected from those nominated by Parties to the Convention and, as appropriate, by intergovernmental organizations, in accordance with guidance provided for this purpose by the Conference of the Parties.

3. The review process shall provide a thorough and comprehensive technical assessment of all aspects of the implementation by a Party of this Protocol. The expert review teams shall prepare a report to the Conference of the Parties serving as the meeting of the Parties to this Protocol, assessing the implementation of the commitments of the Party and identifying any potential problems in, and factors influencing, the fulfilment of commitments. Such reports shall be circulated by the secretariat to all Parties to the Convention. The secretariat shall list those questions of implementation indicated in such reports for further consideration by the Conference of the Parties serving as the meeting of the Parties to this Protocol.

4. The Conference of the Parties serving as the meeting of the Parties to this Protocol shall adopt at its first session, and review periodically thereafter, guidelines for the review of implementation of this Protocol by expert review teams taking into account the relevant decisions of the Conference of the Parties.

5. The Conference of the Parties serving as the meeting of the Parties to this Protocol shall, with the assistance of the Subsidiary Body for Implementation and, as appropriate, the Subsidiary Body for Scientific and Technological Advice, consider:

    (a) The information submitted by Parties under Article 7 and the reports of the expert reviews thereon conducted under this Article; and

    (b) Those questions of implementation listed by the secretariat under paragraph 3 above, as well as any questions raised by Parties.

6. Pursuant to its consideration of the information referred to in paragraph 5 above, the Conference of the Parties serving as the meeting of the Parties to this Protocol shall take decisions on any matter required for the implementation of this Protocol.

Article 9

1. The Conference of the Parties serving as the meeting of the Parties to this Protocol shall periodically review this Protocol in the light of the best available scientific information and assessments on climate change and its impacts, as well as relevant technical, social and economic information. Such reviews shall be coordinated with pertinent reviews under the Convention, in particular those required by Article 4, paragraph 2(d), and Article 7, paragraph

2(a), of the Convention. Based on these reviews, the Conference of the Parties serving as the meeting of the Parties to this Protocol shall take appropriate action.

2. The first review shall take place at the second session of the Conference of the Parties serving as the meeting of the Parties to this Protocol. Further reviews shall take place at regular intervals and in a timely manner.

Article 10

All Parties, taking into account their common but differentiated responsibilities and their specific national and regional development priorities, objectives and circumstances, without introducing any new commitments for Parties not included in Annex I, but reaffirming existing commitments under Article 4, paragraph 1, of the Convention, and continuing to advance the implementation of these commitments in order to achieve sustainable development, taking into account Article 4, paragraphs 3, 5 and 7, of the Convention, shall:

(a) Formulate, where relevant and to the extent possible, cost-effective national and, where appropriate, regional programmes to improve the quality of local emission factors, activity data and/or models which reflect the socio-economic conditions of each Party for the preparation and periodic updating of national inventories of anthropogenic emissions by sources and removals by sinks of all greenhouse gases not controlled by the Montreal Protocol, using comparable methodologies to be agreed upon by the Conference of the Parties, and consistent with the guidelines for the preparation of national communications adopted by the Conference of the Parties;

(b) Formulate, implement, publish and regularly update national and, where appropriate, regional programmes containing measures to mitigate climate change and measures to facilitate adequate adaptation to climate change:

(i) Such programmes would, *inter alia*, concern the energy, transport and industry sectors as well as agriculture, forestry and waste management. Furthermore, adaptation technologies and methods for improving spatial planning would improve adaptation to climate change; and

(ii) Parties included in Annex I shall submit information on action under this Protocol, including national programmes, in accordance with Article 7; and other Parties shall seek to include in their national communications, as appropriate, information on programmes which contain measures that the Party believes contribute to addressing climate change and its adverse impacts, including the abatement of increases in greenhouse gas emissions, and enhancement of and removals by sinks, capacity building and adaptation measures;

(c) Cooperate in the promotion of effective modalities for the development, application and diffusion of, and take all practicable steps to promote, facilitate and finance, as appropriate, the transfer of, or access to, environmentally sound technologies, know-how, practices and processes pertinent to climate change, in particular to developing countries, including the formulation of policies and programmes for the effective transfer of environmentally sound technologies that are publicly owned or in the public domain and the creation of an enabling environment for the private sector, to promote and enhance the transfer of, and access to, environmentally sound technologies;

(d) Cooperate in scientific and technical research and promote the maintenance and the development of systematic observation systems and development of data archives to reduce uncertainties related to the climate system, the adverse impacts of climate change and the economic and social consequences of various response strategies, and promote the development and strengthening of endogenous capacities and capabilities to participate in international and intergovernmental efforts, programmes and networks on research and systematic observation, taking into account Article 5 of the Convention;

(e) Cooperate in and promote at the international level, and, where appropriate, using existing bodies, the development and implementation of education and training programmes, including the strengthening of national capacity building, in particular human and institutional capacities and the exchange or secondment of personnel to train experts in this field, in particular for developing countries, and facilitate at the national level public awareness of, and public access to information on, climate change. Suitable modalities should be developed to implement these activities through the relevant bodies of the Convention, taking into account Article 6 of the Convention;

(f) Include in their national communications information on programmes and activities undertaken pursuant to this Article in accordance with relevant decisions of the Conference of the Parties; and

(g) Give full consideration, in implementing the commitments under this Article, to Article 4, paragraph 8, of the Convention.

Article 11

1. In the implementation of Article 10, Parties shall take into account the provisions of Article 4, paragraphs 4, 5, 7, 8 and 9, of the Convention.

2. In the context of the implementation of Article 4, paragraph 1, of the Convention, in accordance with the provisions of Article 4, paragraph 3, and Article 11 of the Convention, and through the entity or entities entrusted with the operation of the financial mechanism of the Convention, the developed country Parties and other developed Parties included in Annex II to the Convention shall:

(a) Provide new and additional financial resources to meet the agreed full costs incurred by developing country Parties in advancing the implementation of existing commitments under Article 4, paragraph 1(a), of the Convention that are covered in Article 10, subparagraph (a); and

(b) Also provide such financial resources, including for the transfer of technology, needed by the developing country Parties to meet the agreed full incremental costs of advancing the implementation of existing commitments under Article 4, paragraph 1, of the Convention that are covered by Article 10 and that are agreed between a developing country Party and the international entity or entities referred to in Article 11 of the Convention, in accordance with that Article.

The implementation of these existing commitments shall take into account the need for adequacy and predictability in the flow of funds and the importance of appropriate burden sharing among developed country Parties. The guidance to the entity or entities entrusted with the operation of the financial mechanism of the Convention in relevant decisions of the Conference of the Parties, including those agreed before the adoption of this Protocol, shall apply *mutatis mutandis* to the provisions of this paragraph.

3. The developed country Parties and other developed Parties in Annex II to the Convention may also provide, and developing country Parties avail themselves of, financial resources for the implementation of Article 10, through bilateral, regional and other multilateral channels.

Article 12

1. A clean development mechanism is hereby defined.

2. The purpose of the clean development mechanism shall be to assist Parties not included in Annex I in achieving sustainable development and in contributing to the ultimate objective of the Convention, and to assist Parties included in Annex I in achieving compliance with their quantified emission limitation and reduction commitments under Article 3.

3. Under the clean development mechanism:

(a) Parties not included in Annex I will benefit from project activities resulting in certified emission reductions; and

(b) Parties included in Annex I may use the certified emission reductions accruing from such project activities to contribute to compliance with part of their quantified emission limitation and reduction commitments under Article 3, as determined by the Conference of the Parties serving as the meeting of the Parties to this Protocol.

4. The clean development mechanism shall be subject to the authority and guidance of the Conference of the Parties serving as the meeting of the Parties to this Protocol and be supervised by an executive board of the clean development mechanism.

5. Emission reductions resulting from each project activity shall be certified by operational entities to be designated by the Conference of the Parties serving as the meeting of the Parties to this Protocol, on the basis of:

   (a) Voluntary participation approved by each Party involved;

   (b) Real, measurable, and long-term benefits related to the mitigation of climate change; and

   (c) Reductions in emissions that are additional to any that would occur in the absence of the certified project activity.

6. The clean development mechanism shall assist in arranging funding of certified project activities as necessary.

7. The Conference of the Parties serving as the meeting of the Parties to this Protocol shall, at its first session, elaborate modalities and procedures with the objective of ensuring transparency, efficiency and accountability through independent auditing and verification of project activities.

8. The Conference of the Parties serving as the meeting of the Parties to this Protocol shall ensure that a share of the proceeds from certified project activities is used to cover administrative expenses as well as to assist developing country Parties that are particularly vulnerable to the adverse effects of climate change to meet the costs of adaptation.

9. Participation under the clean development mechanism, including in activities mentioned in paragraph 3(a) above and in the acquisition of certified emission reductions, may involve private and/or public entities, and is to be subject to whatever guidance may be provided by the executive board of the clean development mechanism.

10. Certified emission reductions obtained during the period from the year 2000 up to the beginning of the first commitment period can be used to assist in achieving compliance in the first commitment period.

Article 13

1. The Conference of the Parties, the supreme body of the Convention, shall serve as the meeting of the Parties to this Protocol.

2. Parties to the Convention that are not Parties to this Protocol may participate as observers in the proceedings of any session of the Conference of the Parties serving as the meeting of the Parties to this Protocol. When the Conference of the Parties serves as the meeting of the Parties to this Protocol, decisions under this Protocol shall be taken only by those that are Parties to this Protocol.

3. When the Conference of the Parties serves as the meeting of the Parties to this Protocol, any member of the Bureau of the Conference of the Parties representing a Party to the Convention but, at that time, not a Party to this Protocol, shall be replaced by an additional member to be elected by and from amongst the Parties to this Protocol.

4. The Conference of the Parties serving as the meeting of the Parties to this Protocol shall keep under regular review the implementation of this Protocol and shall make, within its mandate, the decisions necessary to promote its effective implementation. It shall perform the functions assigned to it by this Protocol and shall:

   (a) Assess, on the basis of all information made available to it in accordance with the provisions of this Protocol, the implementation of this Protocol by the Parties, the overall effects of the measures taken pursuant to this Protocol, in particular environmental, economic and social effects as well as their cumulative impacts and the extent to which progress towards the objective of the Convention is being achieved;

   (b) Periodically examine the obligations of the Parties under this Protocol, giving due consideration to any reviews required by Article 4, paragraph 2(d), and Article 7, paragraph 2, of the Convention, in the light of the objective of the Convention, the experience gained in its implementation and the evolution of scientific and technological knowledge, and in this respect consider and adopt regular reports on the implementation of this Protocol;

   (c) Promote and facilitate the exchange of information on measures adopted by the Parties to address climate change and its effects, taking into account the differing circumstances, responsibilities and capabilities of the Parties and their respective commitments under this Protocol;

(d) Facilitate, at the request of two or more Parties, the coordination of measures adopted by them to address climate change and its effects, taking into account the differing circumstances, responsibilities and capabilities of the Parties and their respective commitments under this Protocol;

(e) Promote and guide, in accordance with the objective of the Convention and the provisions of this Protocol, and taking fully into account the relevant decisions by the Conference of the Parties, the development and periodic refinement of comparable methodologies for the effective implementation of this Protocol, to be agreed on by the Conference of the Parties serving as the meeting of the Parties to this Protocol;

(f) Make recommendations on any matters necessary for the implementation of this Protocol;

(g) Seek to mobilize additional financial resources in accordance with Article 11, paragraph 2;

(h) Establish such subsidiary bodies as are deemed necessary for the implementation of this Protocol;

(i) Seek and utilize, where appropriate, the services and cooperation of, and information provided by, competent international organizations and intergovernmental and non-governmental bodies; and

(j) Exercise such other functions as may be required for the implementation of this Protocol, and consider any assignment resulting from a decision by the Conference of the Parties.

5. The rules of procedure of the Conference of the Parties and financial procedures applied under the Convention shall be applied *mutatis mutandis* under this Protocol, except as may be otherwise decided by consensus by the Conference of the Parties serving as the meeting of the Parties to this Protocol.

6. The first session of the Conference of the Parties serving as the meeting of the Parties to this Protocol shall be convened by the secretariat in conjunction with the first session of the Conference of the Parties that is scheduled after the date of the entry into force of this Protocol. Subsequent ordinary sessions of the Conference of the Parties serving as the meeting of the Parties to this Protocol shall be held every year and in conjunction with ordinary sessions of the Conference of the Parties, unless otherwise decided by the Conference of the Parties serving as the meeting of the Parties to this Protocol.

7. Extraordinary sessions of the Conference of the Parties serving as the meeting of the Parties to this Protocol shall be held at such other times as may be deemed necessary by the Conference of the Parties serving as the meeting of the Parties to this Protocol, or at the written request of any Party, provided that, within six months of the request being communicated to the Parties by the secretariat, it is supported by at least one third of the Parties.

8. The United Nations, its specialized agencies and the International Atomic Energy Agency, as well as any State member thereof or observers thereto not party to the Convention, may be represented at sessions of the Conference of the Parties serving as the meeting of the Parties to this Protocol as observers. Any body or agency, whether national or international, governmental or non-governmental, which is qualified in matters covered by this Protocol and which has informed the secretariat of its wish to be represented at a session of the Conference of the Parties serving as the meeting of the Parties to this Protocol as an observer, may be so admitted unless at least one third of the Parties present object. The admission and participation of observers shall be subject to the rules of procedure, as referred to in paragraph 5 above.

Article 14

1. The secretariat established by Article 8 of the Convention shall serve as the secretariat of this Protocol.

2. Article 8, paragraph 2, of the Convention on the functions of the secretariat, and Article 8, paragraph 3, of the Convention on arrangements made for the functioning of the secretariat,

shall apply *mutatis mutandis* to this Protocol. The secretariat shall, in addition, exercise the functions assigned to it under this Protocol.

Article 15

1. The Subsidiary Body for Scientific and Technological Advice and the Subsidiary Body for Implementation established by Articles 9 and 10 of the Convention shall serve as, respectively, the Subsidiary Body for Scientific and Technological Advice and the Subsidiary Body for Implementation of this Protocol. The provisions relating to the functioning of these two bodies under the Convention shall apply *mutatis mutandis* to this Protocol. Sessions of the meetings of the Subsidiary Body for Scientific and Technological Advice and the Subsidiary Body for Implementation of this Protocol shall be held in conjunction with the meetings of, respectively, the Subsidiary Body for Scientific and Technological Advice and the Subsidiary Body for Implementation of the Convention.

2. Parties to the Convention that are not Parties to this Protocol may participate as observers in the proceedings of any session of the subsidiary bodies. When the subsidiary bodies serve as the subsidiary bodies of this Protocol, decisions under this Protocol shall be taken only by those that are Parties to this Protocol.

3. When the subsidiary bodies established by Articles 9 and 10 of the Convention exercise their functions with regard to matters concerning this Protocol, any member of the Bureaux of those subsidiary bodies representing a Party to the Convention but, at that time, not a party to this Protocol, shall be replaced by an additional member to be elected by and from amongst the Parties to this Protocol.

Article 16

The Conference of the Parties serving as the meeting of the Parties to this Protocol shall, as soon as practicable, consider the application to this Protocol of, and modify as appropriate, the multilateral consultative process referred to in Article 13 of the Convention, in the light of any relevant decisions that may be taken by the Conference of the Parties. Any multilateral consultative process that may be applied to this Protocol shall operate without prejudice to the procedures and mechanisms established in accordance with Article 18.

Article 17

The Conference of the Parties shall define the relevant principles, modalities, rules and guidelines, in particular for verification, reporting and accountability for emissions trading. The Parties included in Annex B may participate in emissions trading for the purposes of fulfilling their commitments under Article 3. Any such trading shall be supplemental to domestic actions for the purpose of meeting quantified emission limitation and reduction commitments under that Article.

Article 18

The Conference of the Parties serving as the meeting of the Parties to this Protocol shall, at its first session, approve appropriate and effective procedures and mechanisms to determine and to address cases of non-compliance with the provisions of this Protocol, including through the development of an indicative list of consequences, taking into account the cause, type, degree and frequency of non-compliance. Any procedures and mechanisms under this Article entailing binding consequences shall be adopted by means of an amendment to this Protocol.

Article 19

The provisions of Article 14 of the Convention on settlement of disputes shall apply *mutatis mutandis* to this Protocol.

Article 20

1. Any Party may propose amendments to this Protocol.

2. Amendments to this Protocol shall be adopted at an ordinary session of the Conference of the Parties serving as the meeting of the Parties to this Protocol. The text of any proposed amendment to this Protocol shall be communicated to the Parties by the secretariat at least six months before the meeting at which it is proposed for adoption. The secretariat shall also communicate the text of any proposed amendments to the Parties and signatories to the Convention and, for information, to the Depositary.

3. The Parties shall make every effort to reach agreement on any proposed amendment to this Protocol by consensus. If all efforts at consensus have been exhausted, and no agreement reached, the amendment shall as a last resort be adopted by a three-fourths majority vote of the Parties present and voting at the meeting. The adopted amendment shall be communicated by the secretariat to the Depositary, who shall circulate it to all Parties for their acceptance.

4. Instruments of acceptance in respect of an amendment shall be deposited with the Depositary. An amendment adopted in accordance with paragraph 3 above shall enter into force for those Parties having accepted it on the ninetieth day after the date of receipt by the Depositary of an instrument of acceptance by at least three fourths of the Parties to this Protocol.

5. The amendment shall enter into force for any other Party on the ninetieth day after the date on which that Party deposits with the Depositary its instrument of acceptance of the said amendment.

Article 21

1. Annexes to this Protocol shall form an integral part thereof and, unless otherwise expressly provided, a reference to this Protocol constitutes at the same time a reference to any annexes thereto. Any annexes adopted after the entry into force of this Protocol shall be restricted to lists, forms and any other material of a descriptive nature that is of a scientific, technical, procedural or administrative character.

2. Any Party may make proposals for an annex to this Protocol and may propose amendments to annexes to this Protocol.

3. Annexes to this Protocol and amendments to annexes to this Protocol shall be adopted at an ordinary session of the Conference of the Parties serving as the meeting of the Parties to this Protocol. The text of any proposed annex or amendment to an annex shall be communicated to the Parties by the secretariat at least six months before the meeting at which it is proposed for adoption. The secretariat shall also communicate the text of any proposed annex or amendment to an annex to the Parties and signatories to the Convention and, for information, to the Depositary.

4. The Parties shall make every effort to reach agreement on any proposed annex or amendment to an annex by consensus. If all efforts at consensus have been exhausted, and no agreement reached, the annex or amendment to an annex shall as a last resort be adopted by a three-fourths majority vote of the Parties present and voting at the meeting. The adopted annex or amendment to an annex shall be communicated by the secretariat to the Depositary, who shall circulate it to all Parties for their acceptance.

5. An annex, or amendment to an annex other than Annex A or B, that has been adopted in accordance with paragraphs 3 and 4 above shall enter into force for all Parties to this Protocol six months after the date of the communication by the Depositary to such Parties of the adoption of the annex or adoption of the amendment to the annex, except for those Parties that have notified the Depositary, in writing, within that period of their non-acceptance of the annex or amendment to the annex. The annex or amendment to an annex shall enter into force for Parties which withdraw their notification of non-acceptance on the ninetieth day after the date on which withdrawal of such notification has been received by the Depositary.

6. If the adoption of an annex or an amendment to an annex involves an amendment to this Protocol, that annex or amendment to an annex shall not enter into force until such time as the amendment to this Protocol enters into force.

7. Amendments to Annexes A and B to this Protocol shall be adopted and enter into force in accordance with the procedure set out in Article 20, provided that any amendment to Annex B shall be adopted only with the written consent of the Party concerned.

Article 22

1. Each Party shall have one vote, except as provided for in paragraph 2 below.

2. Regional economic integration organizations, in matters within their competence, shall exercise their right to vote with a number of votes equal to the number of their member States that are Parties to this Protocol. Such an organization shall not exercise its right to vote if any of its member States exercises its right, and vice versa.

Article 23

The Secretary-General of the United Nations shall be the Depositary of this Protocol.

Article 24

1. This Protocol shall be open for signature and subject to ratification, acceptance or approval by States and regional economic integration organizations which are Parties to the Convention. It shall be open for signature at United Nations Headquarters in New York from 16 March 1998 to 15 March 1999. This Protocol shall be open for accession from the day after the date on which it is closed for signature. Instruments of ratification, acceptance, approval or accession shall be deposited with the Depositary.

2. Any regional economic integration organization which becomes a Party to this Protocol without any of its member States being a Party shall be bound by all the obligations under this Protocol. In the case of such organizations, one or more of whose member States is a Party to this Protocol, the organization and its member States shall decide on their respective responsibilities for the performance of their obligations under this Protocol. In such cases, the organization and the member States shall not be entitled to exercise rights under this Protocol concurrently.

3. In their instruments of ratification, acceptance, approval or accession, regional economic integration organizations shall declare the extent of their competence with respect to the matters governed by this Protocol. These organizations shall also inform the Depositary, who shall in turn inform the Parties, of any substantial modification in the extent of their competence.

Article 25

1. This Protocol shall enter into force on the ninetieth day after the date on which not less than 55 Parties to the Convention, incorporating Parties included in Annex I which accounted in total for at least 55 per cent of the total carbon dioxide emissions for 1990 of the Parties included in Annex I, have deposited their instruments of ratification, acceptance, approval or accession.

2. For the purposes of this Article, "the total carbon dioxide emissions for 1990 of the Parties included in Annex I" means the amount communicated on or before the date of adoption of this Protocol by the Parties included in Annex I in their first national communications submitted in accordance with Article 12 of the Convention.

3. For each State or regional economic integration organization that ratifies, accepts or approves this Protocol or accedes thereto after the conditions set out in paragraph 1 above for entry into force have been fulfilled, this Protocol shall enter into force on the ninetieth day following the date of deposit of its instrument of ratification, acceptance, approval or accession.

4. For the purposes of this Article, any instrument deposited by a regional economic integration organization shall not be counted as additional to those deposited by States members of the organization.

Article 26

No reservations may be made to this Protocol.

Article 27

1. At any time after three years from the date on which this Protocol has entered into force for a Party, that Party may withdraw from this Protocol by giving written notification to the Depositary.
2. Any such withdrawal shall take effect upon expiry of one year from the date of receipt by the Depositary of the notification of withdrawal, or on such later date as may be specified in the notification of withdrawal.
3. Any Party that withdraws from the Convention shall be considered as also having withdrawn from this Protocol.

Article 28

The original of this Protocol, of which the Arabic, Chinese, English, French, Russian and Spanish texts are equally authentic, shall be deposited with the Secretary-General of the United Nations.

DONE at Kyoto this eleventh day of December one thousand nine hundred and ninety-seven.

IN WITNESS WHEREOF the undersigned, being duly authorized to that effect, have affixed their signatures to this Protocol on the dates indicated.

Annex A

Greenhouse gases

Carbon dioxide (CO_2)

Methane (CH_4)

Nitrous oxide (N_2O)

Hydrofluorocarbons (HFCs)

Perfluorocarbons (PFCs)

Sulphur hexafluoride (SF_6)

Sectors/source categories

Energy
  Fuel combustion
    Energy industries
    Manufacturing industries and construction
    Transport
    Other sectors
    Other
  Fugitive emissions from fuels
    Solid fuels
    Oil and natural gas
    Other
Industrial processes
  Mineral products
  Chemical industry
  Metal production
  Other production
  Production of halocarbons and sulphur hexafluoride
  Consumption of halocarbons and sulphur hexafluoride
  Other
Solvent and other product use
Agriculture
  Enteric fermentation
  Manure management
  Rice cultivation

Agricultural soils
Prescribed burning of savannas
Field burning of agricultural residues
Other

Waste

Solid waste disposal on land
Wastewater handling
Waste incineration
Other

Annex B

Party	Quantified emission limitation or reduction commitment (percentage of base year or period)
Australia	108
Austria	92
Belgium	92
Bulgaria*	92
Canada	94
Croatia*	95
Czech Republic*	92
Denmark	92
Estonia*	92
European Community	92
Finland	92
France	92
Germany	92
Greece	92
Hungary*	94
Iceland	110
Ireland	92
Italy	92
Japan	94
Latvia*	92
Liechtenstein	92
Lithuania*	92
Luxembourg	92
Monaco	92
Netherlands	92
New Zealand	100
Norway	101
Poland*	94
Portugal	92
Romania*	92
Russian Federation*	100
Slovakia*	92
Slovenia*	92
Spain	92
Sweden	92
Switzerland	92
Ukraine*	100
United Kingdom of Great Britain and Northern Ireland	92
United States of America	93

* Countries that are undergoing the process of transition to a market economy.

Bibliography

Books

Alley, Richard B. *The Two-Mile Time Machine: Ice Cores, Abrupt Climate Change, and Our Future*. Princeton, N.J.: Princeton University Press, 2000.

Bailey, Ronald, ed. *Global Warming and Other Eco Myths: How the Environmental Movement Uses False Science to Scare Us to Death*. Rosemont, Calif.: Prima Publishing, 2002.

Burroughs, William James. *Climate Change: A Multidisciplinary Approach*. Cambridge, U.K.: Cambridge University Press, 2001.

Christianson, Gale E. *Greenhouse: The 200-Year Story of Global Warming*. New York: Penguin, 1999.

Cox, John D. *Climate Crash: Abrupt Climate Change and What It Means for Our Future*. Washington, D.C.: Joseph Henry Press, 2005.

Fagan, Brian. *The Long Summer: How Climate Changed Civilization*. New York: Basic Books, 2003.

Gelbspan, Ross. *Boiling Point: How Politicians, Big Oil and Coal, Journalists and Activists Are Fueling the Climate Crisis—And What We Can Do to Avert Disaster*. New York: Basic Books, 2004.

———. *The Heat Is On: The Climate Crisis, the Cover-Up, the Prescription*. New York: Perseus Books, 1998.

Houghton, John. *Global Warming: The Complete Briefing*. Cambridge, U.K.: Cambridge University Press, 2004.

Langholz, Jeffrey, and Kelly Turner. *You Can Prevent Global Warming (and Save Money!)*. Kansas City, Mo.: Andrew McMeel Publishing, 2003.

Leggett, Jeremy. *The Carbon War: Global Warming and the End of the Oil Era*. New York: Routledge, 2001.

Lomberg, Bjorn. *The Skeptical Environmentalist: Measuring the Real State of the World*. Cambridge, U.K.: Cambridge University Press, 2001.

Lynas, Mark. *High Tide: The Truth About Our Climate Crisis*. New York: Picador, 2004.

Maslin, Mark. *Global Warming: Causes, Effects, and the Future*. Stillwater, Minn.: Voyageur Press, 2002.

Mayewski, Paul Andrew, and Frank White. *The Ice Chronicles: The Quest to Understand Global Climate Change*. Lebanon, N.H.: University Press of New England, 2002.

Mendelsohn, Robert, ed. *Global Warming and the American Economy: A Regional Assessment of Climate Change Impacts*. Northampton, Mass.: Edward Elgar Pub., 2001.

Michaels, Patrick J. *Meltdown: The Predictable Distortion of Global Warming by Scientists, Politicians, and the Media*. Washington, D.C.: Cato Institute, 2004.

Motavalli, Jim. *Feeling the Heat: Dispatches from the Frontlines of Climate Change*. New York: Routledge, 2004.

Peterson, Matthew. *Global Warming and Global Politics*. New York: Routledge, 1996.

Philander, S. George. *Is the Temperature Rising? The Uncertain Science of Global Warming*. Princeton, N.J.: Princeton University Press, 2000.

Romm, Joseph J. *The Hype About Hydrogen: Fact and Fiction in the Race to Save the Climate*. Washington, D.C.: Island Press, 2004.

Schneider, Stephen H. *Climate Change Policy: A Survey*. Washington, D.C.: Island Press, 2002.

Spence, Chris. *Global Warming: Personal Solutions for a Healthy Planet*. New York: Palgrave MacMillan, 2005.

Speth, James Gustave. *Red Sky at Morning: America and the Crisis of the Global Environment*. New Haven, Conn.: Yale University Press, 2004.

Victor, David G. *Climate Change: Debating America's Policy Options*. Washington, D.C.: Council on Foreign Relations Press, 2004.

———. *The Collapse of the Kyoto Protocol and the Struggle to Slow Global Warming*. Princeton, N.J.: Princeton University Press, 2001.

Weart, Spencer R. *The Discovery of Global Warming*. Cambridge, Mass.: Harvard University Press, 2004.

Wohlforth, Charles. *The Whale and the Supercomputer: On the Northern Front of Climate Change*. New York: North Point Press, 2004.

Web Sites

This section offers the reader a list of Web sites that can provide more extensive information on climate change and the greenhouse effect, as well as arguments supporting and questioning their scientific merit. These Web sites also include links to other sites that may be of help or interest. Due to the nature of the Internet, the continued existence of a site is never guaranteed, but at the time of this book's publication, all of these Internet addresses were in operation.

Cooler Heads Coalition
www.globalwarming.org

An association of global warming skeptics affiliated with the conservative National Consumer Coalition and the Competitive Enterprise Institute, the Cooler Heads Coalition posts information challenging the scientific basis of climate change and calculating the economic toll the ostensible solutions may engender.

European Commission
europa.eu.int

Part of the policy apparatus of the European Union (EU), the European Commission maintains a Web site detailing the science behind global warming and what the EU and other governmental bodies plan to do to address it.

National Academies Press Collection: Global Warming/Climate Change
www.nap.edu/collections/global_warming

The National Academies Press (NAP) was created by the National Academies, a confederation of several expert panels tasked with advising the federal government on scientific matters. The NAP's Global Warming/Climate Change Collection contains over 150 scientific reports on the subject, all of which can be accessed on this Web site.

National Climatic Data Center

www.ncdc.noaa.gov/oa/climate/climateextremes.html

A division of the U.S. Department of Commerce, the National Climatic Data Center (NCDC) "is the world's largest archive of active weather data." The NCDC's Web site includes numerous scholarly articles chronicling climate change–related phenomena.

Pew Center on Global Climate Change
www.pewclimate.org

The Pew Center on Global Climate Change's mission "is to provide credible information, straight answers, and innovative solutions in the effort to address global climate change." Featured on the site are assorted scientific and policy analyses.

Union of Concerned Scientists
www.ucsusa.org

An independent public interest organization dedicated to a "cleaner, healthier environment and a safer world," the Union of Concerned Scientists (UCS) publishes in-depth global warming–related data and analysis.

United Nations Framework Convention on Climate Change
www.unfccc.int

The United Nations Framework Convention on Climate Change (UNFCCC) was responsible for the drafting of the Kyoto Protocol. The UNFCCC's Web site includes updates on the implementation of the Protocol, as well as links to articles about climate change and climate change policy.

United States Environmental Protection Agency
www.epa.gov

The EPA's official site provides essential information about global warming—as well as other environmental threats—and what can be done to counteract it. A special Global Warming Kids Site provides age-appropriate material for younger researchers.

World Wildlife Fund
www.worldwildlife.org/climate

The World Wildlife Fund (WWF) works to ensure the survival of the planet's endangered species and their habitats. The WWF's Web site features a section detailing the threat posed by global warming and strategies that governments, businesses, families, and individuals can employ to thwart it.

Additional Periodical Articles with Abstracts

More information about global climate change and related subjects can be found in the following articles. Readers who require a more comprehensive selection are advised to consult *Readers' Guide to Periodical Literature, Readers' Guide Abstracts, Education Abstracts, General Science Abstracts, Humanities Abstracts, Social Sciences Abstracts*, and other H. W. Wilson publications.

The Melting Snows of Kilimanjaro. Brook Wilkinson. *Condé Nast Traveler*, v. 39 p105 May 2004.

Tanzania's minister of tourism, Wilkinson reports, has disputed claims by a team of American scientists that one third of Kilimanjaro's ice fields has vanished in the last 12 years and that the remainder will disappear by 2015. Climatologists attribute the snow and ice melt to both global warming and deforestation at the base of the mountain, which decreases the cloud cover and exposes the glaciers to a higher level of solar radiation. The melting of the mountain's ice field could have serious consequences for the people of Tanzania, threatening not just their water supply but the millions of tourist dollars that enter the country every year.

Extreme Weather. Jennifer Vogel. *E: The Environmental Magazine*, v. 16 pp16–18 May/June 2005.

Hurricanes are definitely on the increase, Vogel writes, but it is unclear whether this should be seen as a sign of global warming. Experts maintain that the devastating 2004 hurricane season can only partially be blamed on climate change. According to Ruth Curry, research specialist at Woods Hole Oceanographic Institute, there are several factors that contribute to hurricane formation, including El Niño cycles, upper stratospheric circulation patterns, and the amount of rainfall in the Sahel area of Africa. The 2004 hurricane season is mostly attributed to the alignment of these three crucial elements. The general scientific consensus on climate change and hurricanes is that hurricanes will not necessarily become more frequent but that they will increase in intensity. The writer discusses other weather phenomena that may be affected by climate change.

Carbon Trading: Revving Up. *The Economist*, v. 376 pp 64–65 July 9, 2005.

Prices, participants, and volumes in Europe's carbon market are rising quickly after the introduction of tradable annual allowances for greenhouse gas emissions under the Kyoto treaty that came into effect in February 2005, according to the article. Within the new market-based emissions-trading system for Europe's industrial and energy plants, light polluters are selling some of their surplus allowances to heavier polluters, resulting in an overall reduction of emissions at a lower cost than if each installation had been obliged to meet an individual target. Louis Redshaw of Barclays Capital figures that the allowances to produce one kilowatt-hour of coal-fired power now cost more than the coal itself.

Nuclear Power Is Back—Not a Moment Too Soon. Geoffrey Colvin. *Fortune*, v. 151 p57 May 30, 2005.

Nuclear power is very quietly reemerging in the United States and around the world, and that is a good thing, says Colvin. Three major U.S. utilities have applied for early site permits for new reactors, and interest is growing across Europe. A technology that was unthinkable for decades becomes rehabilitated only through a combination of factors, and by far the most significant is the mainstreaming of the global-warming threat, Colvin argues. Major companies are now supporting greenhouse-gas reduction, and several of the world's top environmentalists now embrace nuclear power, claiming that with the threat of warming, an emission-free power source is important.

Say Hello to Kyoto. Alan Greenblatt. *Governing*, v. 18 p63 September 2005.

State authorities do not share Washington's indifference to the environment, according to Greenblatt. The federal government has shown little interest in dealing with the issue of global warming, and the Kyoto treaty is no longer a relevant issue in the U.S. Senate. Moreover, the big energy bill passed in July 2005 hardly mentioned climate change. In June, however, the U.S. Conference of Mayors adopted a resolution urging Congress and the states to meet the pollution-reduction targets set by the Kyoto treaty and pledging to improve environmental practices in their cities. Individual states are also setting their own targets for emissions reductions.

Meadow's End. Daniel Duane. *Mother Jones*, v. 29 pp64–67 July/August 2004.

Since 1990, University of California–Berkeley professor John Harte has been baking a Rocky Mountain meadow to investigate the effects of global warming. On a slope near Gothic, Colorado, Harte has hung an array of infrared heat lamps across a 100-yard sweep of grasses and blossoms to create real warming, in real time, and in a real ecosystem. He has eschewed historical temperature records, computer models, and thus vulnerability to the claim that his research is conjecture. His results have not been encouraging, Duane reports. Sagebrush is crowding out everything that makes a meadow a meadow—the colors, textures, birds, and bees—and similar experiments elsewhere suggest that mountain meadows will, by the end of the century, become arid expanses. Details of Harte's research are presented, and the implications of his findings are examined.

Boiling Point: Excerpt from *The Boiling Point*. Ross Gelbspan. *The Nation*, v. 279 pp24+ August 16–23, 2004.

Activists and environmentalists must not compromise on climate change, Gelbspan contends. Although climate change is the dominant threat facing human civilization in the 21st century, institutions are doing dangerously little to tackle this problem. The efforts of the fossil-fuel industry and its allies in the Bush administration have contributed to this atmosphere of denial, but large environmental organizations and opposition politicians have also showed a troubling lack of leadership on this crucial challenge. A number of America's leading environmental organizations are promoting limits for future atmospheric carbon levels that are the best they believe they can negotiate. These restrictions may be politically realistic, but they are likely to be environmentally disastrous.

Cooled Down: Global Warming. Steven F. Hayward. *National Review*, v. 57 pp36–38 January 31, 2005.

Climate change is a serious issue, but given the despicable way environmentalists and the Left exploit it, and the inaccurate record of so many past predictions of ecological disaster, profound skepticism is still the reasonable default position, Hayward argues. Indeed, the hyping of the issue has started to backfire on environmentalists, with a spring 2004 Gallup Poll indicating that there is declining public interest in global warming. Moreover, the inability of the scientific community to provide a probability estimate of either a rise in temperature or the effects of such a rise, regionally and globally, illustrates just how limited the world's knowledge of climate actually is. The basic theory of global warming is correct, Hayward claims, but much more work will be needed before the understanding of climate change is comprehensive enough to base trillion-dollar decisions on it.

The Global-Warming God: Research Pertaining to Cause of Hurricanes. Patrick J. Michaels. *National Review*, v. 57 pp24+ October 10, 2005.

For some time, there has been concern that a major hurricane hitting New Orleans could prompt legislation on global warming that would do nothing about tropical cyclones but would damage the American economy. Evidence of this became clear when Robert F. Kennedy Jr. insisted that Hurricane Katrina's severity was connected to President Bush's unwillingness to cap carbon-dioxide emissions, and when Hillary Clinton announced that she hoped to establish a commission to examine the government's response to the storm. Katrina's strength was not affected by global warming, Michaels contends, however, and despite myriad news reports to the contrary, there is no proof that any such warming will lead to more frequent, let alone more intense, tropical cyclones.

Warming the World. Raymond T. Pierrehumbert. *Nature*, v. 432 p677 December 9, 2004.

The writer examines how well Fourier's concept of planetary energy balance measures up against current understanding of the greenhouse effect. Although Fourier had some spectacularly wrong ideas, he did get the essence of the greenhouse effect correct—the principle of energy balance and the asymmetric influence of the atmosphere on incoming light versus outgoing infrared.

Trouble Brews over Contested Trend in Hurricanes. Quirin Schiermeier. *Nature*, v. 435 pp1008–1009 June 23, 2005.

Schiermeier writes of the continuing debate over whether worsening hurricanes are linked to global warming. In January 2005 leading U.S. meteorologist Chris Landsea resigned from the Intergovernmental Panel on Climate Change, complaining that a colleague on the panel, Kevin Trenberth, had supported a link between warming and storms in a press conference. Trenberth has now clarified his views (*Science* 2005:308:1753–1754), arguing that the intensity, if not the frequency, of hurricanes and typhoons will increase as the oceans warm. In contrast, Landsea and colleagues argue in an upcoming *Bulletin of the American Meteorological Society* [November 2005] that no link between greenhouse-gas emissions and hurricane activity has been proved. Trenberth counters that the evidence is being ignored by skeptics.

In a Melting Trend, Less Arctic Ice to Go Around. Andrew C. Revkin. *New York Times*, pp A1+ September 29, 2005.

This summer, the floating ice cap on the Arctic Ocean has shrunk by 500,000 square miles to its smallest size in a century of record keeping, notes Revkin. Experts stated that the development is hardly explainable without global warming and will most likely lead to a rise in ocean temperature.

The Climate of Man—I: Geothermal Evidence from Permafrost in the Alaskan Arctic. Elizabeth Kolbert. *The New Yorker*, v. 81 pp56–71, April 25, 2005.

The theory and hypothetical effects of global warming have become a reality, Kolbert observes in this piece, the first in a series on climate change. In 1979 the National Academy of Sciences undertook its first rigorous study of global warming, through the nine-member Ad Hoc Study Group on Carbon Dioxide and Climate. The panel concluded that if carbon dioxide continued to increase, there was no reason to doubt that climate changes would result and that these changes would not be negligible. Since then, global carbon-dioxide emissions have continued to rise, along with the planet's temperature. Almost every major glacier in the world is shrinking, the oceans are becoming warmer and more acidic, the difference between day and night temperatures is declining, animals are shifting their ranges poleward, and plants are blooming days, sometimes weeks, earlier than they used to.

The Climate of Man—II: Akkadian Empire Collapse. Elizabeth Kolbert. *The New Yorker*, v. 81 pp64–73 May 2, 2005.

In the second article in her climate-change series, Kolbert discusses the role climate has played in the downfall of cultures. The world's first empire was established 4,300 years ago, between the Tigris and Euphrates rivers, by Sargon of Akkad, and it collapsed suddenly after three generations. Yale University archaeologist Harvey Weiss, after discovering a lost city called Tell Leilan, came to believe that the end of the Akkadian empire was the product of a drought so prolonged and severe that it represented an example of climate change. Weiss first published his theory in August 1993, and since then the list of cultures whose demise has been connected to climate change has continued to grow. In each case, what began as a provocative theory has come to seem increasingly compelling as new information has emerged.

The Climate of Man—III: Stabilizing Carbon Dioxide Emissions. Elizabeth Kolbert. *The New Yorker*, v. 81 pp52–63 May 9, 2005.

No matter what people do at this point, global temperatures will continue to increase in the coming decades, due to the gigatons of extra carbon dioxide already circulating in the atmosphere, Kolbert writes in the third piece of her "Climate of Man" series. With over 6 billion people in the world, the risks of this trend are clear. A disruption in monsoon patterns, a change in ocean currents, or a major drought could easily give rise to streams of refugees numbering in the millions. As the effects of global warming become more and more obvious, the question arises as to whether people will react by at last fashioning a global response or by retreating into ever narrower and more destructive patterns of self-interest. It may appear impossible to conceive that a technologically advanced society could choose, in essence, to destroy itself, but that is what is currently happening. The writer discusses recent research on climate change and considers why the Bush administration is ignoring the dangers of global warming.

Storm Warnings: Global Warming and Hurricanes. Elizabeth Kolbert. *The New Yorker*, v. 81 pp 35–36 September 19, 2005.

The United States' consumption of fossil fuels and such catastrophes as Hurricane Katrina are linked, the writer contends. Although hurricanes are, in their details, very complicated, fundamentally they all draw their energy from the warm surface waters of the ocean. It follows that if sea surface temperatures rise, as they have been doing, then the amount of energy available to hurricanes will increase. In general, climate scientists predict that rising carbon dioxide levels will result in an increase in the intensity of hurricanes. With this in mind, President George W. Bush's withdrawl from the Kyoto Protocol in March 2001 could prove costly in terms of future storms.

For the Earth, the Heat Is On. Gretel H. Schueller. *Popular Science*, v. 266 pp52–53 January 2005.

According to Schueller, 2004 brought overwhelming atmospheric evidence that global warming is really occurring and that humans are playing a role in causing it. Across the Northern Hemisphere, meteorologists measured record-setting spring and summer temperatures that triggered unusual behavior among a variety of animal species. In addition, the level of atmospheric carbon dioxide reached a record high, averaging 379 parts per million, representing a jump from 2003 levels that was much greater than the average annual increase of 1.8 parts per million recorded over the past decade. In a striking shift for the Bush administration, a study sponsored by the U.S. government that was released in August supported the view that human action is primarily responsible for global warming.

How Much More Global Warming and Sea Level Rise? William D. Collins, Gerald A. Meehl, and Warren M. Washington. *Science*, v. 307 pp1769–1772 March 18, 2005.

The writers report on two global climate models that show that even if the concentrations of greenhouse gases in the atmosphere had been stabilized in the year 2000, we are already committed to further global warming. Experts predict that by the end of the 21st century mean temperatures will increase by about a half degree and that sea levels will rise an additional 320 percent due to thermal expansion. According to the studies, the projected weakening of the Gulf Stream will not lead to cooling in northern Europe. At any given point in time, even if concentrations are stabilized, there is a commitment to future climate changes that will be greater than those we have already observed.

How Hot Will the Greenhouse World Be? Richard A. Kerr. *Science*, v. 309 p100 July 1, 2005.

Experts are trying to tighten up estimates of how Earth will respond to climate warming, Kerr reports. The sensitivity of new climate models has improved, with predictions that climate sensitivity in response to climate drivers, such as greenhouse gases, is not likely below 1.5°C, though its upper bound is significantly less well constrained. To fully understand Earth's response to climate warming, a better knowledge of clouds and aerosols is needed, as well as an increase in the fidelity of models and more and better records of past climate changes and their drivers.

Facing the Impact of Global Warming. Ginger Pinholster. *Science*, v. 304 pp1921–1922 June 25, 2004.

Experts recently gathered to discuss troubling perspectives on global warming. Researchers attending the conference "Qs and AAAs About Global Climate Change," held on June 15 at the American Association for the Advancement of Science building in Washington, D.C., shared their latest findings and best temperature projections. Authoritative research has revealed that between 1990 and 2100, temperatures will increase between 1.4 and 5.8°C. At the conference, *Science* editor-in-chief Donald Kennedy stated that, although many scientific questions remain regarding climate change, policy makers and the public must take action now.

Behind the Hockey Stick. David Appell. *Scientific American*, v. 292 pp34–35 March 2005.

Appell reports that Michael Mann continues to defend his research on humanity's contribution to global warming. Seven years ago, Mann introduced his so-called hockey stick graph, a plot of the past millennium's temperature showing that temperature remained essentially flat until about 1900, then shot up, like the upturned blade of a hockey stick. Whereas proponents view the plot as a clear indicator that humans are warming the globe, skeptics contend that the climate is undergoing a natural fluctuation not unlike those in past eras. Although he has already defended his research from many unjustified attacks, Mann believes that the attacks will continue because many skeptics obtain funds from petroleum interests. He believes the prediction of regional disruptions will be important to get people to view climate change seriously. He also believes the solution to global warming will be finding an appropriate set of constraints on fossil-fuel emissions that allow the rate of climate change to be slowed down to a level that can be adapted to.

How Did Humans First Alter Global Climate? William F. Ruddiman. *Scientific American*, v. 292 pp46–53 March 2005.

A bold new hypothesis suggests that the farming practices of human ancestors started global warming, Ruddiman writes. According to the scientific consensus, human actions first began to have a warming effect on Earth's climate within the past century. New evidence, however, indicates that concentrations of carbon dioxide began increasing approximately 8,000 years ago, despite natural trends indicating they should have been decreasing, and that methane began to increase in concentration about 3,000 years later. Human activities tied to farming, primarily agricultural deforestation and crop irrigation, may have added the extra carbon dioxide and methane to the atmosphere. Without these surprising increases, current temperatures in northern parts of North America and Europe would be cooler by 3–4°C, and an incipient ice age would likely have started several thousand years ago in parts of northeastern Canada. Sidebars describe how Earth's orbit affects greenhouse gases, the impact of human activity on greenhouse gases, and the relation between disease pandemics and global cooling.

Global Thawing. Carl Pope. *Sierra*, v. 90 pp10–11 May/June 2005.

As the ice caps in the Arctic and Antarctic melt away, there is also a tangible thaw on the issue of energy policy and climate change, Pope notes. The fact that energy policy is back in play means that there is a new opportunity to curb pollution and slow global warming. With the Kyoto Accords being implemented without U.S. participation, increasing numbers of senators and representatives are calling for action on climate change. To support their position, new studies indicate that by 2025, one-quarter of America's energy supplies could come from renewable and agricultural energy sources such as wind, solar, and biofuels. Global warming is at last being seen not as a partisan subject but as a challenge that everyone must and can tackle.

Ice Melt Alert. Janet Larsen. *USA Today*, v. 133 p33 May 2005.

The melting of the Earth's ice cover is accelerating, Larsen oberves. New satellite data show the Arctic region warming more during the 1990s than during the 1980s, and Arctic Sea ice now melting by up to 15 percent per decade. The loss of the Arctic Sea ice could alter ocean circulation patterns and trigger changes in climate patterns worldwide. In addition, Southern Ocean sea ice floating near Antarctica has shrunk by approximately 20 percent since 1950. Recent studies have also shown the worldwide acceleration of glacier melting.

Nuclear Now: How Clean, Green Atomic Energy Can Stop Global Warming. Spencer Reiss and Peter Schwartz. *Wired*, v. 13 pp78+ February 2005.

The burning of coal and other fossil fuels is driving climate change, but nuclear energy provides a sane and practical alternative, Reiss and Schwartz argue. The risks of splitting atoms pale beside the dreadful toll exacted by fossil fuels, and, unlike global warming, radiation containment, waste disposal, and nuclear weapons proliferation are manageable problems. In addition, unlike the usual green alternatives of water, wind, solar, and biomass, nuclear energy is already here and in industrial quantities. Indeed, the latest generation III+ reactors seem to be fuel-efficient, use passive safety technologies, and could be cost-competitive. The writers discuss the effectiveness of a number of renewable energy sources and examine four crucial factors that could help to facilitate the leap from a hydrocarbon to a nuclear era: regulating carbon emissions, revamping the fuel cycle, revitalizing innovation in nuclear technology, and replacing gasoline with hydrogen.

The Irony of Climate. Brian Halweil. *World Watch*, v. 18 pp18–23 March/April 2005.

Climate warming is making the future uncertain for farmers, Halweil reports. Across the world, farmers and climate scientists are finding that generations-old patterns of rainfall and temperature are shifting. Plant scientists think that the most serious threats to agriculture will come from subtle climatic shifts that occur during key periods in crops' life cycles. Most climatologists think that the lack of climate stability will hit farmers in the developing world hardest, but there are measures that farmers can take to cope with climate change. Ultimately, each individual can play a role in combating global warming by buying locally produced food to lessen energy consumption and greenhouse gas emissions.

Arctic Warming Accelerates. Lisa Mastny. *World Watch*, v. 18 p21 January/February 2005.

A new report from the Arctic Climate Impact Assessment warns that the Arctic is now warming at nearly twice the rate of the rest of the world, Mastny notes. According to the report, average winter temperatures in Alaska and western Canada have risen by as much as 4°C in the past 50 years and this warming is largely due to human-caused greenhouse gas emissions during the past century. Ice melt at the North Pole is accelerating because of the warming, which has serious implications for the region's wildlife and people, global sea levels, and overall planetary warming.

Index